BUTTERFIELD'S BYWAY

BUTTERFIELD'S BYWAY

AMERICA'S FIRST OVERLAND MAIL ROUTE ACROSS THE WEST

MELODY GROVES

THE History PRESS

Published by The History Press
Charleston, SC 29403
www.historypress.net

First published 2014

Manufactured in the United States

ISBN 978.1.62619.487.8

Library of Congress CIP data applied for.

For Erin and Haley, always the intrepid explorers

CONTENTS

Prologue: Where It All Began 9

PART I: SETTING THE STAGE—METAPHORICALLY SPEAKING

1. New Mexico Territory—or Is This Arizuma? 17
2. Texas: Ignored but Never Forgotten 21
3. California at the Center 25
4. There's Gold in Them Hills!: The Rush Is On 27

PART II: THE NEED FOR SPEED

5. So We Need to Stitch Together the Coasts—How Do We Do That, Exactly? 33
6. We Have a Railroad, Don't We? 35
7. What About the Route? How Exactly Should We Get Across the Continent? 37
8. Whew! Finally Decided 39
9. Now That We Know Where We're Going, Who Will Lead Us? 41
10. The Rest of the Route: The Oxbow 45
11. Bifurcated Route: Fort Smith to Memphis 47
12. The San Antonio–San Diego Line: A Viable Alternative? 49

PART III: BUILDING THE IMPOSSIBLE

13. Creating the Roads 55
14. Building the Stage Stations 59

15. You're Not Going Anywhere without Water 67
16. Stagecoaches: Cradles on Wheels 69
17. Concord Stagecoaches: The Luxury Liners 71
18. Celerity Stagecoaches: Who Are You Calling a "Mud Wagon"? 73
19. Drivers: The True Kings of the Road 79
20. Horses and Mules: Giddy Up! 81

Part IV: The Trip Itself
21. Opening Day: What a Send-Off! 89
22. From Sea to Sea: What Was It Like, Really? 93
23. A Quick Look at Just How Fast They Traveled 123
24. The Trip Has Ended, but the Comments Are Just Beginning 125
25. To Sum It All Up 127

Part V: The End of an Era
26. Butterfield Impeached/Stepped Down/Reassigned 131
27. The Bascom Affair 133
28. And Now…the Rest of the Story 139
29. The Pony Express 141
30. Telegraph 147
31. Indians 151
32. The Civil War 155

Epilogue 161
Rules of Travel: Stagecoach Decorum 163
Travel Comparison: 150 Years Ago versus Today 165
List of Firsts 167
List of Way Stations 169
Timeline 177
Terms and Slang 179
Bibliography 183
Index 185
About the Author 189

PROLOGUE: WHERE IT ALL BEGAN

Growing up in southern New Mexico, I traipsed up and down and all around the dirt (and later cobblestone) streets of La Mesilla, the tiny village literally in my backyard. As I marched around the plaza, playing on the bandstand where weddings, baptisms, funerals and *quinceañeras* are held, I imagined myself back in the "old days" and fancied being a white-hatted cowboy—or cowgirl! Or maybe I was a friend (or a sweetheart) of Billy the Kid, since he was jailed right here on the plaza's corner. Or maybe…my mind flew with scenarios each and every time I stepped foot on La Mesilla soil. Truth be told, it still does.

And it was here that I discovered the Butterfield Overland Mail Stagecoach Line. I'd heard about it for years, but all I truly knew was that it involved a brightly painted stagecoach and at least four horses pulling it. I even got to ride around the plaza in an original coach!

After reading more and more about Butterfield and talking to generational storytellers of La Mesilla who had secondhand accounts of this stage line, I realized this was something extraordinary.

John Butterfield created the first reliable line of communication by establishing the longest mail route in the world. How special was that?

In Las Cruces, a mile from La Mesilla, I attended a talk by New Mexico's Butterfield expert, George Hackler. "So," I asked, "why was this so super important?" After groaning, shifting his eyes sideways and then knitting his brows at me, he explained patiently that before and especially after the gold rush of 1849, people on the West Coast desired reliable mail service that didn't take four months bobbing over oceans and across the Isthmus of Panama.

I'd heard about the Camino Real, the road stretching from Mexico City to Santa Fe, a north–south route. "Why not use that?" I asked Hackler. He replied that the U.S. government occasionally used it to get messages from Santa Fe down to the Mexican government, but it was unreliable at best. He further explained that while the Santa Fe Trail had opened the West when it first took form in 1821, the eight-hundred-mile route began in Missouri and ended in Santa Fe. That was all well and good, but it didn't help the men and women in California seeking that gold to get letters to and from back home. It wasn't designed as a mail route but rather one of commerce. And it didn't satisfy mail needs in southern New Mexico and El Paso, Texas. In fact, it wasn't until 1854 that mail from Independence started being carried on the Santa Fe Trail at all, taking twenty-five days to make the trip to Santa Fe.

The Pony Express? Not invented until 1860. Other stage lines? They existed but ran only a few hundred miles. What about the train? Mr. Hackler explained that there was no train connecting the east to west; there was also no telegraph, no telephone and, of course, no Internet. In fact, the United States essentially stopped at the Mississippi River and did not resume until the West Coast, with only the states of California (1850) and Oregon (1859) existing. The Great American Desert in between was seen as an area full of "crazed Natives and lawless white people." Who in their right mind would want to travel across, much less live there?

But Californians continued to squawk about the lack of regular mail. Pressure, along with the possibility of untold wealth, was applied on Congress, and it finally authorized the postmaster general to issue a contract.

So, Butterfield WAS a big deal. I'd been right! As a writer of Western novels, my imagination naturally went wild as Mr. Hackler spoke of the planning involved with this 1857–58 undertaking. I was hooked. I learned that John Butterfield, freight line owner and former mayor of Utica, New York, who happened to be a friend of incoming president John Buchanan, won the proposal and bid. Within twelve months, he selected a route; purchased 1,200 horses and 600 mules (each branded with "OM" for Overland Mail) and then distributed them; hired one thousand men as surveyors, conductors, drivers, blacksmiths, etc.; ordered 250 wagons; surveyed 2,800 miles; graded fording sites; opened new roads or improved old ones; procured several thousand tons of hay and fodder; built two hundred way stations every twenty miles; dug one hundred wells; and created the run schedule.

As I researched, read, analyzed and thought about this undertaking, it grew into an incredibly complex Gordian knot of political intrigue, under-the-table shenanigans, financial posturing and all the machinations that come with changing the world. It's taken me years to unravel it. But in the end, the completion and operation of the United States first transcontinental road is close to dumbfounding.

This route was so important, in fact, that on March 30, 2009, President Barack Obama signed congressional legislation authorizing a study in regards to designating it a National Historic Trail. The National Park Service conducted a series of community meetings and a special resource study to determine its fate. It's still pending, but I'm sure they'll see how important this route is.

It's absolutely amazing, I realized. So, as they say in the movies, "You had me at 'giddy-up.'"

PART I
SETTING THE STAGE—METAPHORICALLY SPEAKING

The blast of the stage horn as it rolls through the valleys and over the prairies of the West, cheers and gladdens the heart of the pioneer. As it sounds through the valleys of Santa Clara and San Jose, it sends a thrill of delight to the Californian. He knows that it brings tidings from the hearts and homes he left behind him; it binds him stronger and firmer to his beloved country…The Overland is the most popular institution of the Far West.
—San Francisco Bulletin, *June 13, 1859*

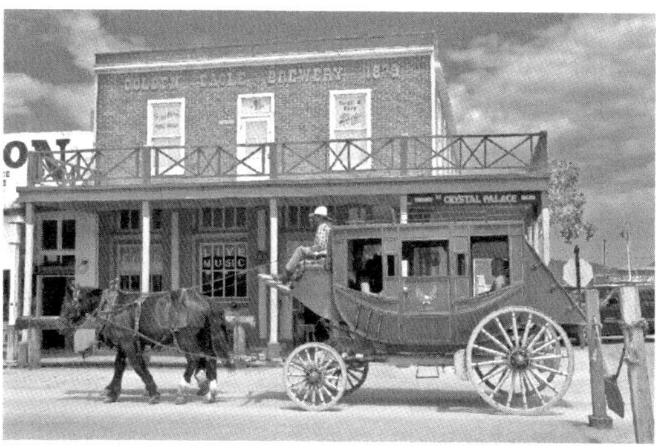

A stagecoach waits outside the Crystal Palace in Tombstone, Arizona, just as it did a century ago. *Photo by Myke Groves.*

"Go west, young man, go west!" These famous words, uttered by twenty-nine-year-old *New York Tribune* founder Horace Greeley, captured venturesome Americans' imaginations. This phrase, uttered around 1850, referred to going all the way west—clear into Ohio Territory. Fortune seekers, people wanting land and/or families longing to

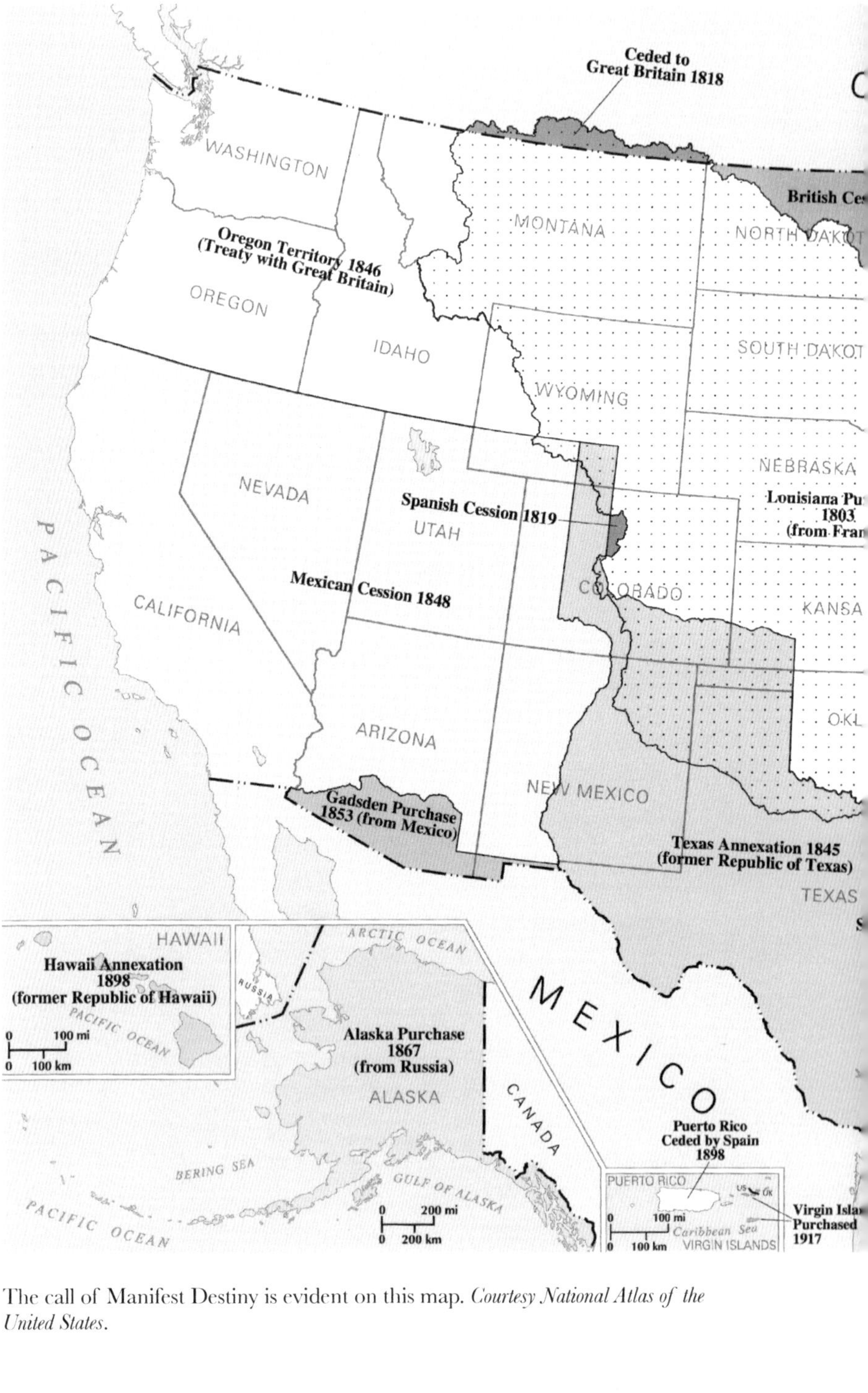

The call of Manifest Destiny is evident on this map. *Courtesy National Atlas of the United States.*

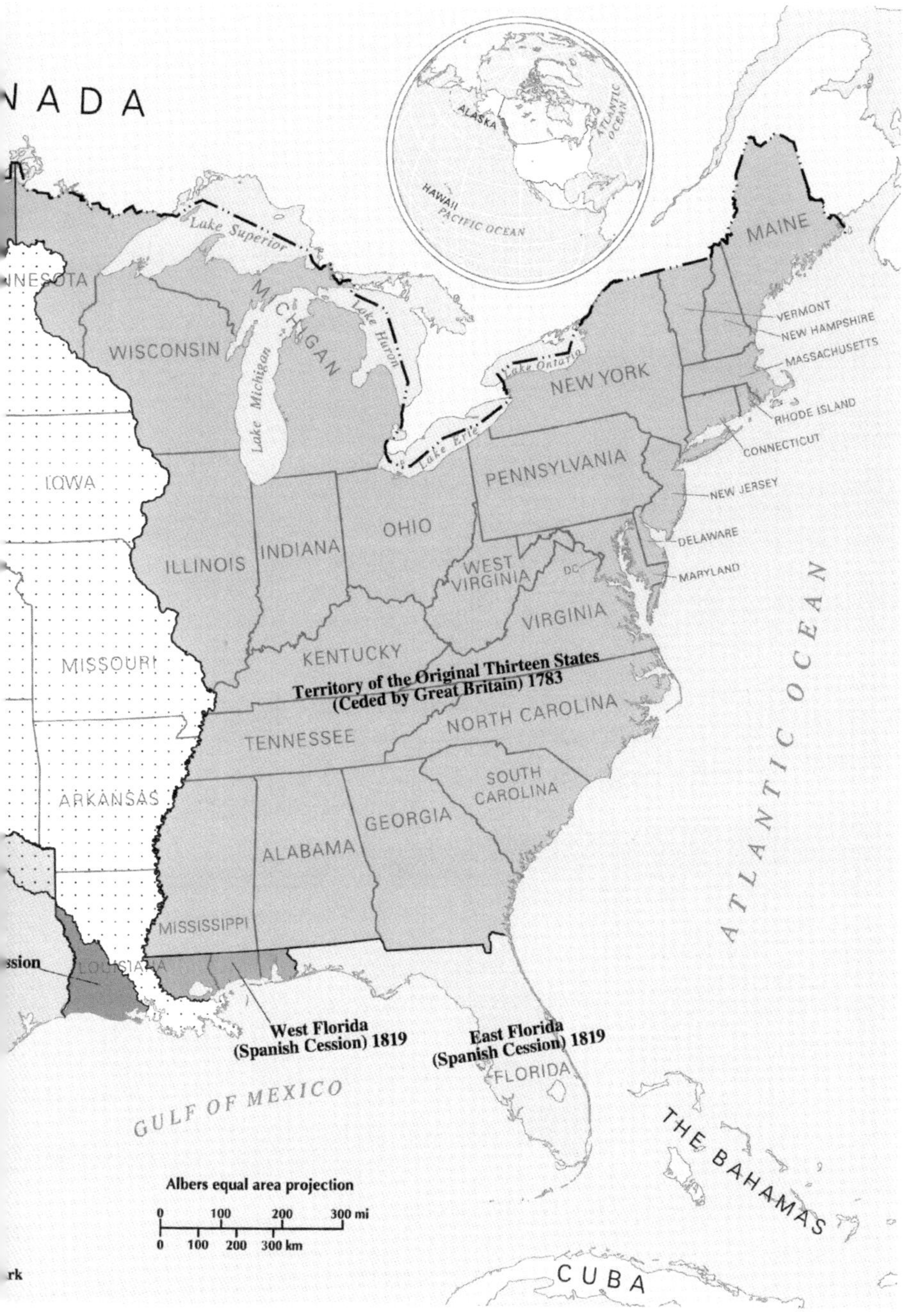
NADA
ALASKA
ATLANTIC OCEAN
HAWAII
PACIFIC OCEAN
Lake Superior
Lake Michigan
Lake Huron
Lake Ontario
Lake Erie
MICHIGAN
WISCONSIN
NESOTA
IOWA
MISSOURI
ARKANSAS
LOUISIANA
MAINE
VERMONT
NEW HAMPSHIRE
MASSACHUSETTS
RHODE ISLAND
CONNECTICUT
NEW YORK
NEW JERSEY
DELAWARE
MARYLAND
DC
PENNSYLVANIA
OHIO
INDIANA
ILLINOIS
WEST VIRGINIA
VIRGINIA
KENTUCKY
Territory of the Original Thirteen States (Ceded by Great Britain) 1783
TENNESSEE
NORTH CAROLINA
SOUTH CAROLINA
GEORGIA
ALABAMA
MISSISSIPPI
ATLANTIC OCEAN
ssion
West Florida (Spanish Cession) 1819
East Florida (Spanish Cession) 1819
FLORIDA
GULF OF MEXICO
THE BAHAMAS
CUBA
Albers equal area projection
0 100 200 300 mi
0 100 200 300 km
rk

forge a better life took his words to heart and headed west. Some even ventured across the Ohio River.

Americans tended to be tenacious in finding ways and means of getting from "Point A" to "Point B," and even then they didn't let the Great American Desert or lack of easy transportation get in their way. They leapfrogged Horace Greeley's concept of the West and found themselves on the Pacific coast. The Spanish and Mexicans were already in California, but that didn't stop Americans from making room for themselves. America was a continent in the process of being crossed and consolidated one way or another. The dream of transcontinental transportation grew into a substantial part of the land-hungry way of thinking.

The United States of the mid-1800s was one immense unsettled area from present-day Maine west to the Mississippi River and from Minnesota to Florida and Texas. For a western boundary, other than the Mississippi, the Rio Grande meandered at will. It would be years before dams along the way kept the wandering riverbeds in check.

Only California (1850) and Oregon (1859) existed as western states, sprinkled with a few widely scattered settlements. Texas (1845), located in the extreme South, was not at all sure exactly where its western boundary was. The Rio Grande tended to meander. In reality, the United States was east of the ninety-seventh meridian, including, for the most part, the inhabited sections of Texas, its settlements spread more than one hundred miles apart.

In response to the pressure, in March 1853, Congress included $150,000 in the army budget for the purpose of surveying routes to the west. The routes were to carry mail and small packages. Eight were considered, although only four emerged as potential winners.

Still, Manifest Destiny (a term coined in 1844) ruled. An unnamed traveler in the 1850s remarked, "Congregate 100 Americans beyond the settlements and they immediately lay out a city, form a state constitution and apply for admission to the Union, while 25 of them become candidates for the United States Senate."

Chapter 1

NEW MEXICO TERRITORY—OR IS THIS ARIZUMA?

So while it was clear that speedy mail and passenger service across the continent was necessary and even essential if the United States was going to fulfill its Manifest Destiny, how to go about doing that exactly was the issue. A railroad seemed the perfect solution except for one tiny detail: the United States didn't own the land stretching from Southern California to Texas—an easier route without mountains and winter snows. Come to find out, it owned land just north of where the train was supposed to run.

A bit of backstory might clear things up.

The Mexican-American War (1846–48) ended with the signing of the Treaty of Guadalupe Hidalgo on February 2, 1848. It ceded to the United States 1.2 million square miles of the present-day Southwest, encompassing all previously held Mexican territory north of the Rio Grande and Gila River. It provided the U.S. boundary for Texas, all of California, Arizona, New Mexico, Utah, Nevada and parts of Wyoming and Colorado. The United States paid Mexico $15 million plus another $3.25 million to pay off American claims against the Mexican government. This treaty was declared "official" by President James K. Polk on July 4, 1848.

So it would appear that the United States had its land on which to build the Pacific Railroad. Not so fast. Before someone noticed a "glitch" on the map, and with the lure of gold in its eyes, the United States decided to build that railroad, with its southern connection, from Texas to California. Secretary of War Jefferson Davis (future president of the Confederate States of America) and James Gadsden planned to use the railroad to transport southerners to the California gold fields. Oh yes, they had tried once, but they failed to divide California into two states—the southern half allowing slavery.

The "glitch," as it turns out, was big. The treaty specified that the Rio Grande boundary would veer west eight miles north of El Paso. However, it was based on an 1847 copy of a twenty-five-year-old map. More current surveys revealed that El Paso was thirty-six miles farther south and one hundred miles farther west than the map showed. Mexico favored the map, but the United States trusted the results of the survey. The disputed territory involved a few thousand square miles and about three thousand residents. Bordering the Rio Grande, the area essential for the construction of the transcontinental railroad, it consisted of flat desert land measuring about fifty miles north to south by two hundred miles east to west.

It was a pretty major "glitch" by anyone's standards—a big oops. Never one to give up, Davis sent Gadsden to Mexico City in 1852 to negotiate the purchase of land west of the Rio Grande, south of the Gila River and east of the Colorado River—the part originally "left off." This 30-million-acre Pennsylvania-sized slice of land eventually sold for $0.33 an acre, or $10 million (close to $280 million today). The Gadsden Purchase, the last major territorial acquisition in the lower United States, known also as the Treaty of La Mesilla, was signed in 1854.

It was so unpopular on both sides of the border that the deal caused the Mexican dictator, Antonio López de Santa Anna, to be overthrown and forced into exile. Mexicans felt he betrayed their country when, in fact, he was trying to amass money to rebuild an army in defense against the United States. In the United States, many people thought the price paid (again) was excessive for such a desolate parcel of sand. Antislavery senators saw it as an underhanded ploy to gain more slavery territory. No one was happy. Orator Daniel Webster said, "What do we want with this region of savages and wild beasts, of shifting sands and whirlwinds of dust, of cactus and prairie dogs? To what use could we ever put those endless mountain ranges? What could we do with the western coast of three thousand miles, rock ground, cheerless and uninviting? Improve communications with the west? No, gentlemen. I for one will not vote one penny for such an enterprise. The whole western country is not worth it."

Indeed, the Gadsden Purchase was done strictly for an all-weather route for the future railroad. Captain Randolph B. Marcy had previously staked out a route from Fort Smith, Arkansas, to El Paso. In response, the people of Fort Smith petitioned Jefferson Davis, then secretary of war, to accept

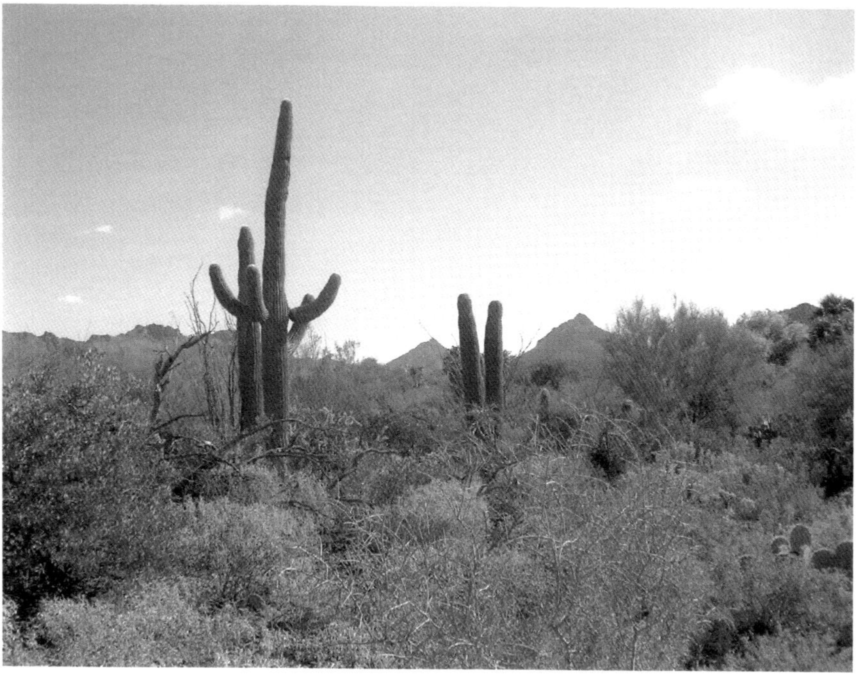

Towering saguaro cactus help define the Southwest's Sonoran Desert. *Public domain.*

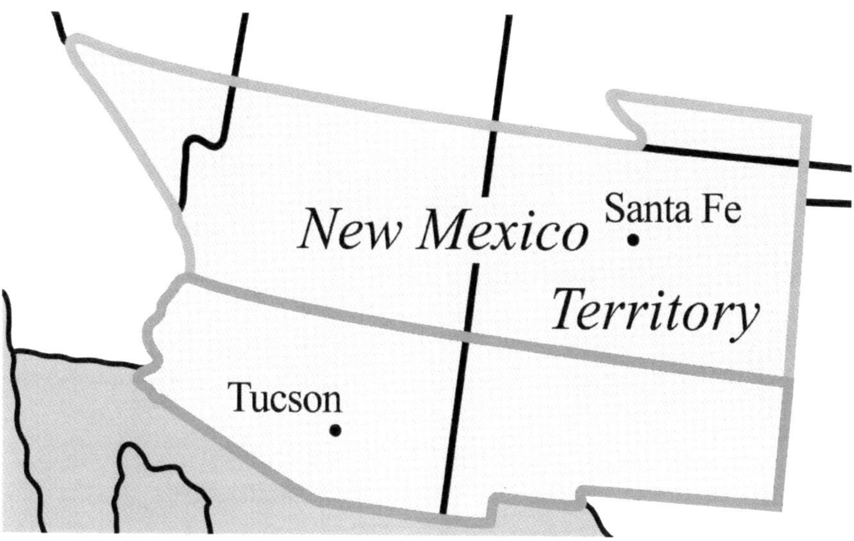

In 1860, Arizona and New Mexico Territory were divided horizontally rather than vertically. *Public domain.*

the route. Davis refused, citing his attempt not to favor the South. Now there's some irony!

However, after the treaty was signed, a concerted effort—along with a series of petitions to Washington—proposed naming the region that stretched from La Mesilla (New Mexico) to Arizona City (Yuma) as Arizuma, Arizonia, Gadsdenia, Pimeria (after the Pima Indians) or Arizona. The last one stuck.

The largest settlement between Fort Smith, Arkansas, and San Francisco, California, was La Mesilla, New Mexico Territory. Close to two thousand called it home when the Overland Stage opened up in 1858. Then it grew to three thousand. An 1850 census shows less than six thousand in all of New Mexico, including Santa Fe. Arizona City boasted twenty permanent settlers in 1857, and Franklin, Texas (present-day El Paso), consisted of "…20 single-story adobe buildings, most of which were vacant."

However, the sale of the land helped establish a new and permanent boundary—the Rio Grande. By 1877, the Southern Pacific Railroad tracks stretched from California to Arizona. But it wasn't until 1883 that the line coming from Texas linked with the Southern Pacific. Once completed, however, the railroad extended from New Orleans to Los Angeles.

Chapter 2

TEXAS: IGNORED BUT NEVER FORGOTTEN

Another major player in the U.S. quest for Manifest Destiny and a southern railroad route was Texas. People from various parts of the country brought different and even clannish ways with them when they migrated west. It was basically New Englanders who went by ship to settle Oregon and Northern California, while southerners migrated to Texas and Southern California. Those who would become Texans literally walked into what was then Mexico.

Starting in about 1790, *Norteamericanos* migrated south toward Texas, which was still officially owned by Spain but controlled by Mexico. Mexican officials not only permitted Anglo-American settlers to enter the Texas wilderness but encouraged it as well. The hope was that the men and women would civilize the region and, in return, become loyal, happy Mexican citizens.

And come they did. Stephen F. Austin led the colonization effort starting along the Brazos River in 1821. At that time, his colony numbered about five hundred, which, considering that there were only four thousand people in all of Texas, was huge. Within ten years, predictable clashes between Mexican and American cultures ensued and increased. By 1830, Austin's colony numbered five thousand, compared to all of Texas at twenty thousand. Conflicts worsened in the 1830s when Mexico won its independence from Spain and President Antonio López de Santa Anna assumed dictatorial powers.

While Santa Anna built his empire, Stephen F. Austin, along with Sam Houston, Davy Crockett, William Travis and James Bowie, led attacks on

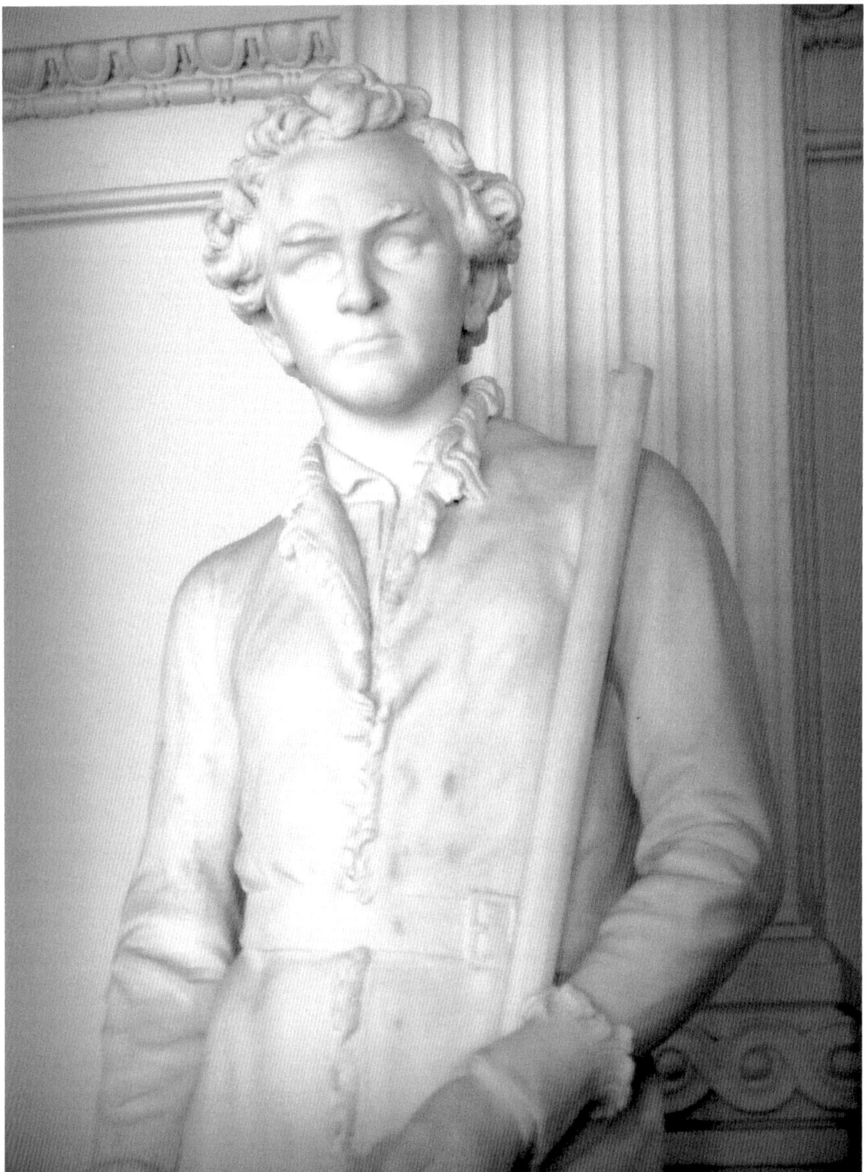

Stephen F. Austin (1793–1836) was known as the "Father of Texas." This marble sculpture was created by Elisabet Ney and stands at the Texas State Capitol. *Photo by Kafziel.*

the Mexican Army. Battles at San Antonio, the Alamo and Gonzalez showed the strength of both forces. Finally victorious, Sam Houston led his army into San Jacinto, giving Texas the right to declare independence on April 21, 1836.

In 1836, Sam Houston became the first president of the Republic of Texas. Immediately, officials pushed for annexation to the United States. In 1837, a formal offer for annexation was made, but antislavery forces in the United States objected strongly to the addition of an enormous slave state. The offer was withdrawn.

Houston, elected to a second term in 1841, secretly made overtures to England for acquisition. After finding out and fearing that England would act, the U.S. Congress, with a push from President John Tyler, responded with a joint resolution for annexation on February 28, 1845. It allowed Texas to enter as a state, skipping the territorial stage. Meanwhile, in Texas, the newly elected president, Dr. Anson B. Jones, considered offers from England and France. Jones delayed making a final decision and conducted polls among citizens and the Texas congress. The decision? Unanimous recommendation of annexation by the United States. Happily, on October 13, 1845, Texas voters approved annexation, and on December 29, 1845, President James Polk signed the act making Texas the twenty-eighth state.

A mere six years later, however, Texas withdrew from the Union, siding with the Confederates during the "War of Northern Aggression."

Even before statehood, mail service in Texas was especially undependable, unreliable and hopelessly irregular. There were no real roads, and even the trails were difficult to maneuver. The situation did not improve after 1845. Now a state, the new Americans clamored for reliable mail service and demanded equal treatment. With California on the brink of statehood, many Texans demanded to know why mail was going to Californians and not them.

Why, indeed. It was a good question, and one answered more politically than socially. The Anglo-American population of Texas in 1850 was 212,592, and by 1858, the only settlements were few and far between. In fact, there was nothing between Franklin (present-day El Paso) and Sherman, located in northeast Texas near present-day Oklahoma. The span—with nothing to break the monotony—is approximately 342 miles. So, with the obvious deficiency in population—not to mention the lack of gold or silver fields—Texas was primarily ignored.

Chapter 3

CALIFORNIA AT THE CENTER

A big player in the U.S. expansion was California. But in 1846, Northern California was a distant and forgotten outpost of Mexican territory. So far from anywhere, its populace felt they owed allegiance to no one. If pushed, they preferred to be annexed and pay taxes to either the United States or England—which one didn't really matter. For its part, the United States was interested in expanding to the Pacific Ocean (part of Manifest Destiny), and the San Francisco Bay area was considered prime. It offered a deep bay where ships from important countries could dock; economies would boom. And if handled correctly, it would be slave free.

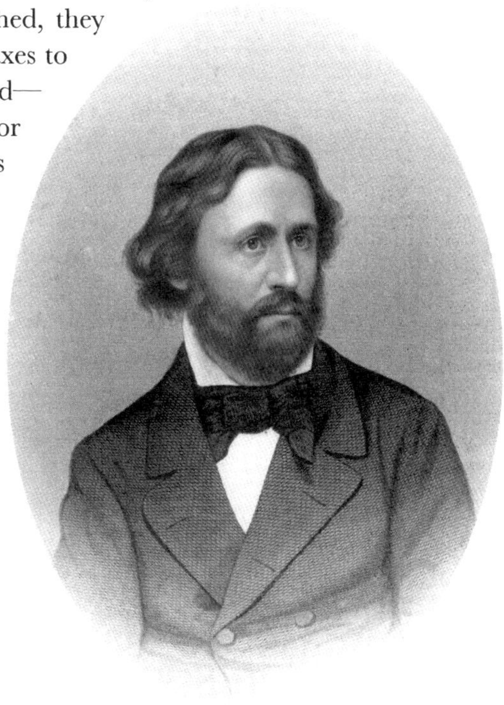

Explorer John C. Fremont (1813–1890), the first antislavery Republican presidential candidate. Historians call him the "Great Pathfinder." *Photo by Matthew Brady, 1856.*

In Northern California, Switzerland-born John A. Sutter constructed a fort on fifty thousand acres that Mexico had granted him in 1841. The fort turned into a way station welcoming newcomers and served as a staging area for the conquest of California. Topographer John C. Frémont led an exploring party there. In June 1846, Frémont and a band of American settlers in the Sacramento Valley staged an uprising against the Mexican government and invaded a settlement in Sonoma. The insurgents raised their Bear Flag and declared California to be free and independent of Mexico. General Stephen Watts Kearny, leader of the Army of the West, and Commodore Robert F. Stockton of the U.S. Pacific Fleet, joined forces in 1847 and occupied Los Angeles, ending hostilities. A treaty with Mexico was signed on January 13, 1847, in what is now North Hollywood.

California would be annexed by the United States thirteen months later with the signing of the Treaty of Guadalupe Hidalgo.

Chapter 4

THERE'S GOLD IN THEM HILLS!: THE RUSH IS ON

"I found gold!" When that elusive metal was discovered at Sutter's Mill on January 24, 1848, no one could envision how those three little words would change not only the United States but the entire world as well. Men and women flocked to California by way of any mode of transportation available. Some rode horseback from the East, while others bought wagons and headed out on the Oregon Trail, which was established in 1841, and then down a complicated network of roads. Still others jumped ship from the East Coast across the Gulf of Mexico to Panama, where they then took burros across the Isthmus to the Pacific.

Whichever way they came, the trip was hard—and, for many, not really worth it. Historians have called the gold rush the largest voluntary migration in human history.

The *California Star* published news of the gold discovery on April 1, 1848, and when copies arrived in New York in August, gold fever was more like a passing fancy. However, it wasn't until the U.S. War Department put on display a tea caddy containing over fourteen pounds of California gold in Washington that people sat up and took notice. *That* caught everyone's attention.

The discovery flamed a worldwide mania, luring miners from all points of the globe. By January 1, 1849, every ship passage from New York to the West Coast was sold out for months in advance. About 16,000 adventuresome souls rounded Cape Horn in 1849, while over 4,600 sailed to Panama, crossed the Isthmus and then tried to buy a ship ride up to San Francisco.

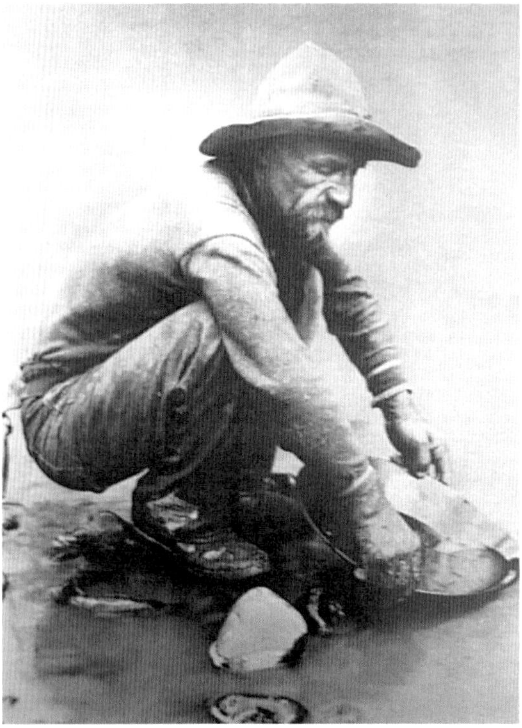

Left: This prospector peers into the silt of California's American River while panning for that elusive gold. *Photo by L.C. McClure, 1850.*

Below: The actual spot near Sutter's Mill where James W. Marshall first found gold is now part of Marshall Gold Historic Park in Coloma, California. *Photo by Bobak Ha'Eri.*

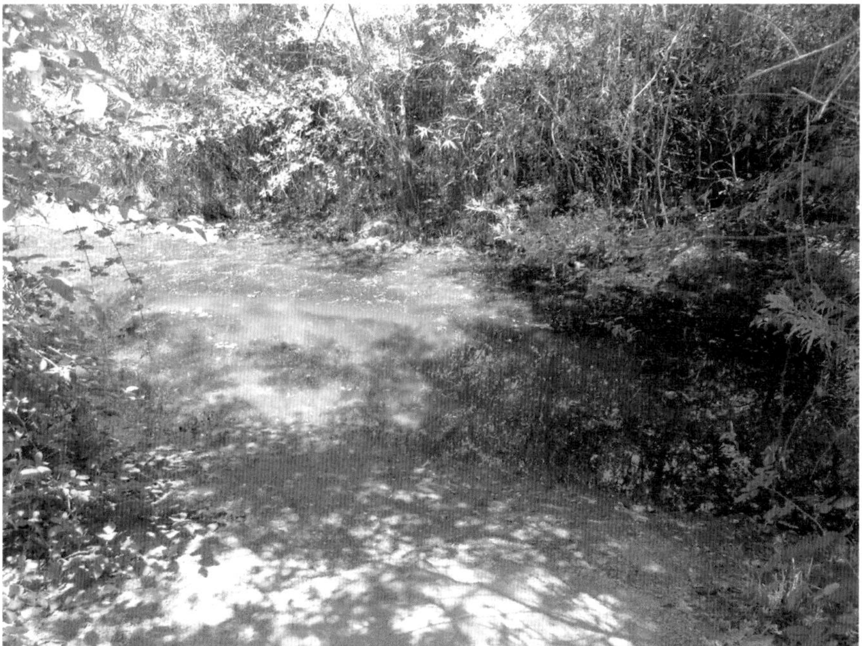

Crossing the Isthmus was no easy feat. From the East Coast, they hired native dugouts and paddled their way down the dangerous Chagres River for twenty-five miles. From there, they rode mules for another twenty miles on the old Spanish trails that had not been maintained for fifty years. The trip across Panama took four to eight days in good weather. Once on the other side, men and women then caught a ship sailing north to San Francisco. In fact, in January 1849, seven hundred people met the first steamship on the West Coast of Central and North America in Panama City. With all scrambling for passage, only four hundred were able to board. As an interesting aside, when the ship reached San Francisco, everyone except a handful of crew deserted the ship, leaving it idling in the bay for four months. Finally, they found enough coal and a new (read expensive) crew.

As many as thirty-five thousand crossed the south pass in Wyoming and went on to California in 1849. This massive migration surge stripped the trails of forage and littered the wagon road with thousands of dead oxen and mules. Indians did little to bother the people, but cholera swept through the unsanitary camps killing hundreds, if not thousands.

Was it worth it? That depends on who's telling the story. For the men and women who were already in California, chances were they made some money. Woe be it to those late getting the news—many ended up working for somebody else or leaving empty-handed. Those who stayed faced disease, famine and a lawless society. Ever resilient, they built lives in this newly acquired territory by farming rich land, establishing new businesses and re-creating themselves.

It has been mentioned by more than a couple historians that the discoveries of gold and silver in places like Nevada and Utah were due to forty-niners being elbowed out of California.

Ever vigilant and ready to expand, the United States spotted the potential and acted. It ratified the territory into statehood in 1850. The population of California had boomed to over 300,000 (compared to 2,400 in 1840), and while they busily went about living, they craved mail and newspapers from "back home." Now with statehood and gold on their side, the demand for mail service made more of an impact on Washington. Surprisingly—or perhaps not—the government listened.

The fact is, they'd been listening for years as a few strong-willed Americans hollered and complained. But back in the spring of 1842, Francisco Lopez washed out flakes of gold in San Feliciano Canyon near the tiny village of Los Angeles. The first gold shipment out of California was made by Abel Stearns, a Los Angeles merchant of questionable repute, in the spring of

1843. He shipped twenty ounces of gold to the mint in Philadelphia. This piqued the interest of the United States enough to send out a series of surveyors for a route to California.

A mail service to the Pacific coast was established on March 3, 1847, with ships that could be converted to war use. However, the route included sailing up and down both coasts and travel over the Isthmus of Panama (by oxen wagons and then boats). Mail still took between three to four months. So, with Americans in California, New Mexico Territory and Texas demanding mail service, Congress finally took notice.

PART II
THE NEED FOR SPEED

Chapter 5

SO WE NEED TO STITCH TOGETHER THE COASTS—HOW DO WE DO THAT, EXACTLY?

Even before and especially after the Mexican-American War of 1846–48, the involvement of the federal government was an outcome of the attitudes and interests of those who were the loudest. The idea of mail to California was as much geographical as it was political. Both were equally difficult to maneuver.

By 1857, the population of the United States was scattered all across North America. The eastern third of Texas had been settled; Arkansas and Missouri were also settled. Iowa, to some degree, sported settlements up to the Missouri River as far as Council Bluffs. Even Minnesota had small villages, consisting mainly of trappers. The northern half of Wisconsin was unsettled. It was clear the western boundary was nebulous at best. No one truly knew where the country ended.

In between the coasts, a few small Spanish and Mexican villages dotted the Rio Grande Valley above and below Santa Fe. And a settlement, appearing somewhat like an island, grew in an area near Salt Lake. That village was about eight hundred miles west of any "civilized" area and about seven hundred miles from the sparse settlements along the Pacific coast. The fact was, not many people lived in between the East and West Coasts.

California in the early 1850s was composed of a series of well-established settlements in the San Joaquin Valley and four scattered villages stretching south to San Diego. And all along the Pacific coast from Canada to Mexico were strung a total of eight towns, the largest in the San Joaquin Valley area.

Although there were few inhabitants west of the Missouri River, they were loud and heard in Washington. "Mail to California!" was the cry. Some sort of continental mail service was a sure thing, but the hassle over where the route would go seemed monumental. Surveyors fanned out seeking the best east–west route. By the mid-1850s, California already had a number of "short haul" stage lines operating in various parts of the state, adding up to approximately two thousand miles within the state itself.

The Santa Fe Trail, having been well used since 1821, was a logical choice for transcontinental mail—except for the facts that it ended in Santa Fe and the Rocky Mountains and Sierra Nevada to the west made travel difficult. If that wasn't enough, it was still eight hundred miles from Santa Fe to the Pacific coast. Still, they were willing to try it out, and mail via the Santa Fe Trail began as a regular service in 1854. But it took twenty-five days to get from Independence, Missouri, to Santa Fe. The voices of Californians and Texans grew louder.

While emigrants demanded reliable mail service, the establishment of an overland mail route was actually encouraged by the railroad interests as a trailblazer for the railroads themselves. If a route worked for the stagecoach, it would probably work for a railroad as well.

In 1847, mail service by ship was approved by Congress, which provided for the building of four naval steamships that could be converted for war duty. The bill itself made it the task of the secretary of the navy to contract with the Post Office Department to carry the mail to Panama twice weekly each way. Ships were to be constructed under navy supervision and contracted out to civilians. Curiously, they would be commanded by naval officers.

The Pacific Mail Steamship Company was quite notorious for its monopoly on ocean service and its poor service. The postage rate in March 1848 was forty cents per letter, which equates to about nine dollars today. It dropped to thirty-four cents in 1851, and by 1855, it was ten cents.

Steamboats chugged up and down the Mississippi delivering mail and packages north and south while at the same time flatboats steamed up the Sacramento River to the gold fields in California. All of this ship activity ran mostly north–south, but there were still thousands of east–west miles uncovered by mail service. People complained.

Three major events from the past three decades collided, setting the stage for western acceleration. First, the Mormon migration to the Great Salt Lake Basin created a need for roads, military protection and mail service to Utah. Second, as discussed, the discovery of gold at Sutter's Mill inflamed the nation. Third, the culmination of the Mexican-American War and the addition of the Gadsden Purchase gave land and political clout to a largely population-deficient area.

Chapter 6

WE HAVE A RAILROAD, DON'T WE?

Railroads were already an established institution in the East. By the early 1800s, they had changed the style of transportation. The American railroad mania began with the Baltimore and Ohio Railroad in 1828. No longer did travelers have to endure hours on horseback, an open-air wagon or buggy or the cramped quarters of a stagecoach. Roads were full of potholes, stones and mud. There was nothing easy about traveling before railcars.

Train travel offered a relatively headache-free ride, and in some cases, a dining car was available. And travel time was cut considerably. Even in the South, small railroads linked cotton-growing communities with oceanic or river ports. These links, however, were short and did not connect—which became a real problem during the Civil War.

By 1850, nine hundred miles of line had been built, and by 1860, railroads in the North and Midwest connected almost every major city. So what was to keep railroads from connecting the East and West Coasts? Nothing but the Rocky Mountains and the Great American Desert, as described earlier. That and a lack of funding.

The thundercloud of impending war loomed over politicians' heads. It was no secret that the Southerners were unhappy with the way Washington treated them and that they spoke of seceding. Thus, it was no secret that people with money, the New York financiers, shied away from investing millions of dollars into a project that may not ever come to fruition. That would be especially true as some of the moneyed folks were Southern sympathizers. And while there had been surveys to find the best railroad route across the continent, all of that was put on the back burner until things became more settled.

The first transcontinental railroad would not be completed until May 10, 1869, when the gold spike was pounded in place at Promontory Point, Utah. But it would still be twelve years before the second transcontinental railroad became a reality. It did, indeed, follow the southern route through the Gadsden Purchase, being completed on March 8, 1881. A silver spike finished the job in Deming, New Mexico Territory. The Southern Pacific Railroad was no longer a dream.

Chapter 7
WHAT ABOUT THE ROUTE? HOW EXACTLY SHOULD WE GET ACROSS THE CONTINENT?

Newspaper editors, commercial and industrial sections in the Northeast and the emigrants in the West all set the country afire. However, the question of states' right, with the slavery issue at the crux, circled over politicians' heads like vultures. Added to all of this were the concept of Manifest Destiny and the natural American spirit of exploration and expansion. What a time!

Political posturing is nothing new. It's gone on for centuries, and the years from 1840 through the 1860s were no exception. Eight possible routes for the first transcontinental stage/mail route were proposed, and only four were taken seriously.

Before, during and after the Treaty of Guadalupe Hidalgo, a transcontinental mail route became the prime topic of conversations, editorials and political discussions. Push came to shove, hearings were held and a series of plans emerged. Four routes were offered: two northern routes, one across the plains to Albuquerque and then west and another more southerly route. Proponents for each route backed their favorite with editorials and a whole lot of political glad-handing and proper posturing.

Route number one—the most northerly—commenced at St. Louis; continued on to Independence, Missouri; ran along the north fork of the Platte River through the south pass of the Rockies and down to Salt Lake City; and then continued almost due west to San Francisco.

Route number two commenced at Springfield, Missouri, and then followed the Canadian River to Albuquerque, New Mexico Territory. It then turned at the Mojave River and ran roughly through the Tulare and San Joaquin Valleys to San Jose and then up to San Francisco. Favored by Congress and the financiers of East Coast companies, this "central" route would be the eventual choice for railroad development. It was also a chief contender for the famous first overland route, but it posed one major problem: mountains. The Rockies, as well as the Sierra Nevadas, stood in the way. During the winter, feet of snow piled up, delaying the mail and making travel treacherous at best. However, John Butterfield and his contemporaries rallied in support of this route.

Route number three, called a "southern" route, commenced at St. Louis and then went on to Springfield and Fort Smith, Arkansas. From there, it went southward to Sherman, Texas, and then west to El Paso, where it crossed the Rio Grande and headed directly west to Fort Yuma, California. After crossing Imperial Valley, it headed north to San Francisco.

Route number four was generally known as the "Jim Birch route"—named after the man who founded the "Jackass Mail" line that ran from San Antonio to San Diego. This route made no allowances for mail going north of San Antonio.

Chapter 8
WHEW! FINALLY DECIDED

And the winner is…route number three!

Known as the Oxbow Route, the most southerly of the four proposed routes was adopted by Tennessee native and postmaster general Aaron V. Brown. Chosen mainly because it was an all-year trail, this southern route through the Chihuahua, Sonoran and Mojave Deserts also had political advantages. At the time, the mail was viewed as an instrument of settlement (like freeways and rapid transit today) and thus a means to gain Southern sympathizers. Besides, Aaron Brown had been a senator and governor of Tennessee.

Senator Toombs of Georgia said the following about the chosen route: "The Butterfield route was not selected as a southern route or a northern route, but as a United States route, upon the cheapest, best and nearest way of getting the mails through."

For many northerners, the choice of the southern route added insult to injury. The *Sacramento Union*, home to a city entirely missed by the southern route, called it a "foul wrong," "an outrage upon the majority of people of the state," "a Panama route by land" and "an overland route to Mexico, a military road to Texas, and an emigrant route to Arizona." The East and North felt the same way. The *New York Press* called it the "horseshoe and ox-bow route," while the *Chicago Tribune* condemned it as "one of the greatest swindles ever perpetrated upon the country by the slave-holders."

Postmaster General Aaron V. Brown answered the criticism in his annual report in December 1857, citing that they had chosen "a route that could

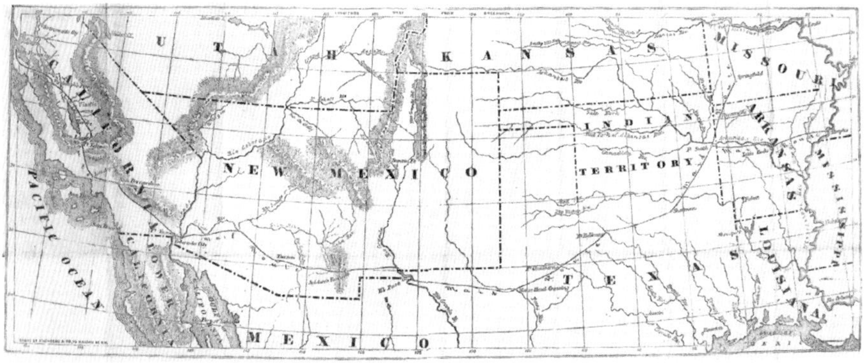

Butterfield had his work cut out for him when he set out to cross this wide expanse of mostly unexplored land. *Public domain.*

be found safe, comfortable, and certain during every season of the year, as well as for the transportation of the mails and for the accommodation of emigrants and the future location of a railroad to the Pacific." It was felt this route would serve three valuable purposes: (1) it would open dealings with Mexico, (2) it would help both nations commercially in times of peace and (3) it would furnish a highway for U.S. troops in case of war. This last reasoning was recently seen in modern days with the establishment of our interstate highway system, its purpose moving military machinery expediently.

Chapter 9
NOW THAT WE KNOW WHERE WE'RE GOING, WHO WILL LEAD US?

In the midst of all this clamor of effort and opinion, one man stood out—a man who was able to translate the chaos into results: John Warren Butterfield.

There are times, history says, where the right man for the job turns out to be the only man for the job. Such was the case with John Butterfield, whose work ethic was two-fold: face tasks and then solve them.

A busy man in the field of stagecoach lines and transportation, Butterfield was able to attract—and keep—good workers. A successful businessman, he didn't seek glory and glamour but was much more interested in results. He'd had little formal education, but he made up for it with his natural organizational and managerial talents. Born in Berne, New York, in 1801 on a farm located directly on a stagecoach route west of Albany, the constant rattle of those big vehicles sparked his interest in stagecoaching. From his home in Upstate New York, he saw Fulton's first steamboat come up the Hudson River to Albany, the first telegraph lines and the first steam locomotive. All this influenced the boy who grew up to impact these inventions.

By the age of nineteen, Butterfield was driving spring coaches for the livery stable firm of Thorpe and Sprague out of Albany. He subsequently moved to Utica and married Malinda Harriet Baker in 1822. The couple would eventually have nine children—four daughters and five sons. Hired by Jason Parker and Company, Butterfield drove a hack between canal boat landings and taverns, picking up business as he went, much like today's taxis. Needing to make ends meet and provide for his growing family, he bought a small two-seat carriage and moonlighted at night. Eventually, he and his wife opened a boardinghouse.

In 1825, Butterfield assumed management of Jason Parker and Company and continued to add to his own inventory of equipment and livestock. This enterprise soon grew into eastern stagecoach lines, packet boats and railroads. Since he had put down roots in Utica, he actively campaigned for public works to keep his adopted hometown ahead by seeking trolley service, railroad connections and street and road construction and repair. He expanded his business until he had a controlling interest in most of the important mail and passenger lines in northwestern New York State.

By the time he was fifty-six, Butterfield had accumulated a comfortable fortune, and it was widely known that no man in the country knew more about the ins and outs of horse-drawn transportation. He had an incredible memory and was a natural-born leader, admired because of his basic generosity and genuine interest in public benefit. He was scrupulously fair.

In 1846, he organized the Butterfield and Wasson Express Company, based on the simple idea that you could charge a lot for guaranteeing the pickup and delivery of valuable property or papers (thus the term "express"). The more valuable the cargo, the higher the charge—regardless of size and weight.

Others, however, had the same idea. Two companies, Wells & Company and Livingston & Fargo, competed with Butterfield's business. Following many conversations, Butterfield convinced Wells and Fargo to join him in a combined new company, American Express, in 1850. And yes, it's the same company that operates worldwide today with the Concord coach logo of Wells Fargo. Wells, Fargo and Butterfield all founded today's American Express/Wells Fargo Bank. Butterfield became the first president and director of that company, retaining the title until his death in 1869.

A broad-shouldered man with a prominent nose, heavy eyebrows and dark hair, John Butterfield was a direct descendant of Benjamin Butterfield, who brought his family to the bay colony in 1638. Settling in New York, the family prospered. John Butterfield's grandfather saw action in the American Revolution, while others in his family participated in the French and Indian Wars. Butterfield's third son, Daniel, served as a brigadier general in the Union army during the Civil War, was awarded the Congressional Medal of Honor and wrote the words for the American salutatory song "Taps."

Definitely a busy man, John Butterfield was the originator of the horse-drawn streetcar and was active in raising capital to construct plank roads. These roads, constructed of plank boards laid over the roadway on log foundations, made transportation easier and faster. Wagons loaded with merchandise and agricultural products, along with the animals pulling them, were not slowed or stopped by mud. Like many inventions, the plank road was a progressive idea and enhanced local travel. By 1855, several hundred miles of these roads had been built, and three hundred plank-road companies were chartered by New York State.

Butterfield was also a director of the Utica City National Bank and mayor of that city from 1855 to 1856. Besides being deeply religious, he was active in New York State politics and had many political connections. Butterfield counted President James Buchanan as a friend.

At a time when he should have retired to enjoy comfortable life with a comfortable fortune behind him, Butterfield instead chose to commit himself to the most outstanding achievement of his career: the Overland Mail Company. Well aware that the days of long-distance stagecoaching were numbered, the iron horse pressing farther into the country, he pushed on. Determined more than ever to bring mail to the California gold fields, he worked tirelessly. On September 16, 1857, Butterfield, along with his associates William Dinsmore, William Fargo, James Gardner, Marcus Kinyon, Alexander Holland, D.N. Barney, E.P. Williams, Hugh Crocker, Giles Hawley, David Moulton and Hamilton Spencer, formed the Overland Mail Company. John Butterfield was elected president, while Dinsmore was elected vice-president and a member of the Wells Fargo Express Company was elected treasurer. They agreed to a six-year contract, at $600,000 per annum, for a semiweekly mail service between St. Louis and San Francisco. By terms of the post office appropriation bill, the company had to commence service within one year of signing.

And indeed, it did. One year to the day later, the stage carrying the mail for the East left San Francisco on September 15, 1858, and the westbound mail left St. Louis on September 16, 1858.

Chapter 10

THE REST OF THE ROUTE: THE OXBOW

Mail would be taken by train 160 miles from St. Louis to Tipton, Missouri. From this point, it would be loaded on the stage and follow a near southerly route to Springfield, Missouri; over the Ozarks to Fayetteville; and then on to Fort Smith.

The route entered Choctaw country just west at Skullyville, the former capital of the Choctaw Nation, following an old Indian trail out of Fort Smith that had been used in 1837 by the Chickasaws. Interestingly, in October 1858, the Choctaw Nation began to charge a toll to any and all travelers crossing their land. For each four-wheeled wagon or other vehicle drawn by four or more horses, mules or oxen, the toll was fifty cents. The Overland Mail Company made no issue and paid the toll. As there were twelve stations located 13 to 19 miles apart in the Indian Nation, the company felt it in everyone's best interest to work with the Choctaw. The entire route across the nation was 197 miles.

After running through the Choctaw Nation, the route crossed the Red River at Colbert's Ferry. Unlike the bigger, more substantial ferries in the East, this was a raft pushed across the shallow river by poles in the hands of Benjamin F. Colbert's sturdy slaves. Colbert, by the way, usually charged $1.25 for a four-horse hitch, but Overland Mail crossed free of charge, as he felt it would increase the travel of others. It was reported that with the surge of westward traffic, he was able to net about $1,000.00 per year—a considerable sum (about $27,000 today). From Colbert's, the route continued southwest across an almost entirely uninhabited region of northern Texas to Fort Chadbourne.

On leaving there, it ran along the Pecos River, through Guadalupe Pass and on to Franklin (El Paso). The route then headed northwest to La Mesilla and then due west to Tucson across 320 miles of waterless desert. From Tucson, it ran north to Picacho Peak, turned west, continued on to Fort Yuma (then near San Diego), crossed the mountains at Warner's Pass and continued to Los Angeles. North up the central valley of California, it turned west through Pacheco Pass and then north to San Francisco.

The route totaled over 2,795 miles. The first run of the Butterfield stages carried the mail and one passenger, Waterman Lily Ormsby Jr. The trip took twenty-three days and twenty-three hours—the shortest time ever.

The route was the longest stage line in the world and the first truly transcontinental road, stitching the United States together. For 1,100 miles in the middle, it crossed the Chihuahua, Sonoran, Colorado and Mojave Deserts. The trip across Texas alone took eight days.

Chapter 11

BIFURCATED ROUTE: FORT SMITH TO MEMPHIS

Who says the government doesn't listen? A second route from Fort Smith to Memphis was tacked on. That posturing made sense because the postmaster general, Aaron V. Brown, was a true southerner. He wanted his citizens to get mail service, and by darn, he'd make sure they did. This section was one of the items that had been part of the original compromise in order to gain the southern vote.

The official route had three termini: St. Louis, Memphis and San Francisco. However, the bifurcated route left a lot to be desired. What had started as a political gesture grew into a major sideline issue. The company had originally planned to transport mail and passengers by water—down the Mississippi and up the Arkansas River to Little Rock and Fort Smith. However, it was discovered that boats were unable to go beyond Little Rock. Butterfield subleased to a stage company operating between Fort Smith and Des Arc on the White River, from which point the mail went overland to Memphis and passengers were transferred to boats.

The subcontractors failed to deliver, and trouble began. The second mail from St. Louis was delayed thirty-seven hours from Fort Smith while waiting for the Memphis stage, and a *San Francisco Bulletin* reporter had to wait three days at the same point. Neglect mounted right along with complaints. The *Bulletin* reporter wrote:

> *The only place of any slight importance, bearing the semblance of a village, was Dardanelle, which is nearly equidistant between Fort Smith and Des*

> *Arc on the Arkansas River. We were most unceremoniously turned out of the coach by the driver and delivered into the charge of the ferryman, who took the mail-bags on his shoulder, and his lantern in hand, told us to follow him to his boat at the ferry landing, about one mile distant up the river. There was no remedy for this unexpected tramp; so, placing our blankets on our backs, and valise in hand, the passengers proceeded to accompany him, through a torrent of rain, up the riverbank and across the stream in his boat…We all got soaking wet by the time we reached the coach on the opposite bank and three of our party were considerably used up, next day, from the effects of the drenching.*

They did finally reach Memphis, of which the same reporter noted, "The city is about two miles long and one mile wide, with a suburban belt quite closely settled…"

With those kind of reports and threats of contract cancellation, John Butterfield himself took over the route and furnished the service required. With his help, both east and westbound schedules were fulfilled, and while people weren't totally satisfied with the service, they didn't grumble as loudly.

The first eastbound passenger, Goddard Bailey, a special agent for the U.S. Postal Department, stepped into the stage in San Francisco and got off in Memphis. He reported, "In conclusion, I have to report that with the exceptions mentioned, the company has faithfully complied with all the conditions of the contract."

Chapter 12

THE SAN ANTONIO–SAN DIEGO LINE: A VIABLE ALTERNATIVE?

With the cry of Manifest Destiny heard coast to coast and in every saloon, statehouse and newspaper, it is no wonder stage lines pushed westward from the more established East Coast. Once people arrived, they needed to get from here to there. And it would be only a matter of time before mail and people traversed the Great American Desert. One such line, shortly before the Overland did the impossible, was the San Diego–San Antonio Line. Its history is well intermingled with that of Butterfield's Overland and that of California, New Mexico and Texas.

The financial powerhouses of the day were mostly, if not all, New York businessmen who ventured into stagecoaching and the express businesses. Their names are linked so closely that at times it's difficult to ascertain who was at the helm of what. And John Butterfield was directly in the mix. His fortunes are mixed with the story of the San Antonio–San Diego Line.

Henry Skillman won a government contract to carry mail from Santa Fe, New Mexico, to San Antonio, Texas, beginning in 1851 and running until 1854. Skillman's selection was justified in an editorial on September 22, 1851, in which the *Daily National Intelligencer* stated he was "well known on the frontier, and highly recommended as a man of energy, intelligence and firmness…There are few men well qualified to perform with punctuality and success the difficult duty he has undertaken."

At first, Skillman carted mail on horseback but eventually switched to a spring wagon and coaches. Despite a good beginning, the endeavor was soon mired in trouble. West of San Antonio until the Rio Grande villages—a span of five hundred miles—there were no relay stations. And from La Mesilla and Santa Fe, there was only a scattering of military posts and tiny villages. Various bands of Indian tribes found the coach and the herd of spare animals being dragged along too tempting. In 1852, the second westbound mail left San Antonio on time but failed to reach its destination. A heavily armed search party finally found a pile of cold ashes surrounded by charred wheels. A battered hat with a bullet hole in its crown lay nearby.

Early in 1854, the postmaster general called for new bids for the Santa Fe–San Antonio route. Surprisingly, David Wasson, a Pennsylvania native who had never been "out West," won the bid to continue the contract. Naturally, Skillman was stunned (he had bid $50,000, Wasson $16,750), acknowledging that pushing the mail through the desert was costly, dangerous and difficult beyond Wasson's experience. That knowledge proved correct, as Wasson failed to meet his contractual requirements, and Skillman was there to pick up and move the mail—not totally legally. He and a couple co-conspirators took it upon themselves to continue the run, with or without Wasson or his representatives.

Several months later, Wasson arrived in San Antonio with a story of being without proper funds. He confessed he could neither repay the loans nor cover the debts that his creditor had incurred in his name. Enter George Giddings, a San Antonio merchant who had loaned Wasson money to start up the line. When Wasson defaulted, Giddings had the mail contract signed over to him. Giddings and Henry Skillman were good friends, so they pooled their resources to save the Mail Company.

To complicate matters, in 1857, there were enough Americans in California and the Southwest to develop a bit of political power. The government did not renew Giddings's contract and awarded James E. Birch a temporary mail contract for semimonthly service from San Antonio to San Diego, allowing thirty days for a one-way trip. This line became known as the "Jackass Mail."

James Birch, a New Englander born in 1827, had moved to California at age nineteen for the express purpose of hauling miners and supplies to the gold diggings. By the end of his first year, he had established a stage line running daily from Sacramento to Sutter's Mill. Soon he operated both stage routes and telegraph lines to Nevada City clear to Stockton from his headquarters in Sacramento. By 1855, he was married and influential enough to step down from the company's helm and begin planning a new enterprise.

Birch wanted to create a transcontinental stage line. He thought, planned, researched and even purchased supplies and materials for the endeavor.

California and Texas supported him. Birch was ready when Postmaster General Aaron V. Brown called for bids. Eight other contractors submitted, their prices ranging from $520,000 to $1 million. Birch's bid was $600,000 for semiweekly service on the exact route specified by Brown. Convinced he would win, he relaxed until learning that a bid of the same amount was submitted by a company formed by the union of the Adams, American, National and Wells Fargo express companies, headed by John W. Butterfield.

On June 22, 1857, Postmaster Brown awarded Birch a consolation prize. For $149,800 per year, he was to commence a semimonthly service between San Antonio and San Diego, thirty days per trip. Brown awarded a six-year contract for the semiweekly Mississippi–San Francisco route to Butterfield for $600,000 per year.

Birch was unhappy to say the least, having led a personal crusade to establish an overland mail service from the Mississippi River area to San Francisco. So it came as a great shock when John W. Butterfield won the bid. Nevertheless, Birch immediately set about readying his new express line.

Many believed that Birch's contract was just an experiment to fill the gap and test the feasibility of the concept and the route. For twelve months before Butterfield's Overland Mail took the reins, the San Antonio–San Diego moved mail, freight and people over the New Mexico/Arizona desert.

For the eastern section of the route, Birch's chief lieutenant, Isaiah Woods, sought out the aid of George Giddings to help organize the line. Shortly afterward, Woods hired Henry Skillman to serve as conductor of the first westward mail run in Arizona. Two days after Skillman's departure, Woods proposed a merger of the two companies, citing the use of combined stock to help keep costs down. Giddings thought several days and finally chose to merge. Thus the San Antonio–San Diego line was born.

In an odd twist of fate, James Birch drowned on September 11, 1857, with the sinking of a ship, the *Central America*, off the South Carolina coast. He was returning from San Diego by sea, headed to New York with a counter proposal to his temporary contract in hand. When the ship went down in the hurricane, it took with it $1.6 million of California gold. Survivors told of losing Birch out of a life raft.

When Butterfield took over, Giddings had over four hundred mules and twenty-five passenger wagons working, but he was under-financed, and the Indians were destroying most of his property along the route. The U.S. Post Office Department took away his contract for the route from Franklin/El Paso to Fort Yuma and gave it to Butterfield. It also gave away his El Paso–Santa Fe contract. This left the San Antonio–San Diego with only two widely separated runs—San Antonio to El Paso and Fort Yuma to San Diego. This contract, however, was upgraded to weekly service.

Technically, the San Antonio–San Diego line shared with Butterfield nine hundred miles of the same trail at the same time. From its inception, Birch's line suffered from Indian pillages—stolen stock and raided stations along the entire road on which Butterfield operated. It was relentless Indian incursions that caused the demise of the San Antonio–San Diego line. The final degradation occurred at Stein's Peak when George Giddings's brother, John, and several others were killed by Cochise-led Chiricahua Indians.

By 1860, the route was modified by discontinuing the line from San Diego to Fort Yuma and changing service in Texas. When Texas seceded from the Union, all mail lines stopped for a time, including the San Antonio–San Diego.

PART III
BUILDING THE IMPOSSIBLE

Chapter 13

CREATING THE ROADS

Although Butterfield had been given a year to have 2,800 miles of road built, the task seemed impossible. Existing roads in the eastern section, being more established, were easier to navigate than the largely uninhabited West, since no adequate roads were there. True, the government had given an earlier stab at surveying and road building, and at least ten different surveys for possible routes across Texas alone were done between 1847 and 1853. But nothing permanent was decided or done.

By 1855, California's mail was delivered by steamship, the first service having been put into effect in 1847 while California was still part of Mexico. California presented a petition of seventy-five thousand signatures—most of questionable validity—demanding more reliable mail service. In 1856, Congress appropriated $600,000 for the "improvement of overland transportation." This money was to be spent improving roads west of El Paso and for rebuilding the road through the south pass in Wyoming. $200,000 was designated for the West, and the rest went to Wyoming. Butterfield understandably was able to use little of that appropriated money.

Building the roads was no easy feat. The West was deemed an inhospitable, vast wasteland, and using money to construct pathways across it was considered by some to be crazy and even downright undemocratic. But Butterfield and his crew persevered.

For hundreds of those miles, no road existed and, in fact, had never existed. Butterfield surveyors followed animal, trapper, military, explorer and emigrant trails. Over the years, those latter trails had been expanded to include carts and wagons, which caused the roads to be a bit wider and flatter and made them easier to navigate. The people making them did not have transcontinental travel in mind, but Butterfield and his surveyors did. And where there were no roads, Butterfield's men graded, scraped and leveled. By government instructions, the road was to be eighteen feet wide on the straightaways and twenty-five feet wide on the curves.

However, in many places along the route, Butterfield did more actual road construction than the government agents who were contracted to do it. The company's personnel checked out the actual ground over which the stages would run. It was one thing to draw a line on a piece of paper and another to actually drive over it.

Oddly, the contract did not include any funding for road building, but Butterfield added road building and maintenance crew to his personnel roster. The Overland Mail traveled some of the distance on existing roads, some laid out by the Army of the West and others created by travelers and local citizens. At the very best, these roads were two lanes through the sand, rocks or grass.

In Arizona and New Mexico, the route lay almost entirely within the Gadsden Purchase. An engineer's report on the construction stated that there were two kinds of improvements. The first was the reduction of surface grades to make ascension easy. The second type of improvement was the increase and/or enhancement of watering places. The formation of six new wells or watering stations reduced the great distance between camps to twenty-seven miles.

The trail through Texas, which was one quarter of the entire route, was especially difficult to tame. In the northern part of the state, southwest of Castle Rock, the road went through a steep-walled canyon known as Mountain Pass. The route from Fort Chadbourne to the Rio Grande was five hundred miles of cactus, mesquite and chaparral, with an unending series of rivers and creeks, all of which were normally dry but could turn into raging floods following rain in the mountains and hills to the west.

In 1859, a year into the Overland Mail run, Butterfield's engineers built the trail more southwestwardly because of the difficulty with water supplies. They put in a ferry at the Pecos River and dropped the route to Comanche Springs near Fort Stockton.

Not all funding was provided by Congress. For example, the people of Sherman, Texas, discovered that Butterfield intended to bypass their small town by eight miles. To prove that their village would be of greater advantage than the other route, they used their own funds and set to work

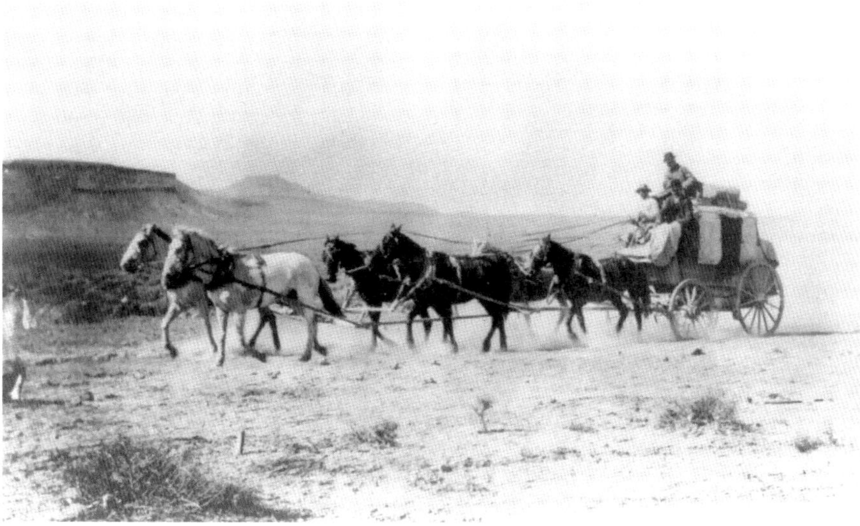

The stagecoach driver was required to turn each pair of animals separately so that the team did not become tangled up and overturn the stage. *Public domain.*

on both the roads and the Overland Mail Company. It worked. Jack County, Texas, opened almost thirty new miles less than two weeks before the first stage rumbled over it.

Young County, Texas, appropriated all of $150 to cut a road through the chaparral and clay banks.

Certainly, by the time the Overland Mail sent its first stagecoaches east and west, sections of the route had been somewhat cleared and many of the extra steep grades leveled or bypassed.

As an aside, it would be several years after Butterfield shut down in 1861 that surveyors went out and discovered that, as a result of confusion over the U.S./Mexico boundary, a series of Overland Mail stage stations south of Fort Yuma, California, were actually in Mexico. Butterfield's first through passenger, newspaper reporter Waterman L. Ormsby, mentioned this in his letters to his editors, but no one seemed overly concerned.

The Overland was recognized as "king of the road." One blast from his bugle, and the stage conductor would have other wagons and animals cleared off the trail. The blast, accompanied by the loud shout of "Make way for the United States mail!" scattered other travelers. Everyone knew the Overland was king. Even today, some of our modern highways are literally on top of the old Butterfield trail.

Chapter 14
BUILDING THE STAGE STATIONS

Having passable roads was of utmost importance, but a permanent place to water and change the animals also proved a major concern. Additionally, a place where drivers could pass the reins to relief drivers had to be built. Butterfield and his advisors decided to build two types of stage stations: the swing station and the home station.

Up until the Overland Company's invention of these "rest areas," stage lines made it a practice of taking everything they would need along with them. They herded extra horses or mules and then stopped and camped out along the trail whenever and wherever was considered convenient. Travelers camped overnight and continued the journey the following day.

But this wouldn't work for what Butterfield had in mind, so he ordered the building of two hundred stage stations. Set an average of fifteen to twenty miles apart, these buildings were made out of local stone, wood or adobe (a mixture of mud and straw). Unique in the history of stagecoaching, the Overland was the first to establish the concept of home and swing stations, which are much like the truck stops and rest areas located along today's interstate system.

Throughout New Mexico and Arizona, the stations were large square enclosures with adobe walls. Supplies of hay, grain and water had to be hauled long distances. However, in Los Angeles, a splendid brick depot was used. It consisted of an office, a blacksmith shop, stables and sheds. On both coasts, established ranches and farmhouses served as stations.

Home stations included permanent construction of corrals, outbuildings and maintenance/repair shops where harness makers and blacksmiths,

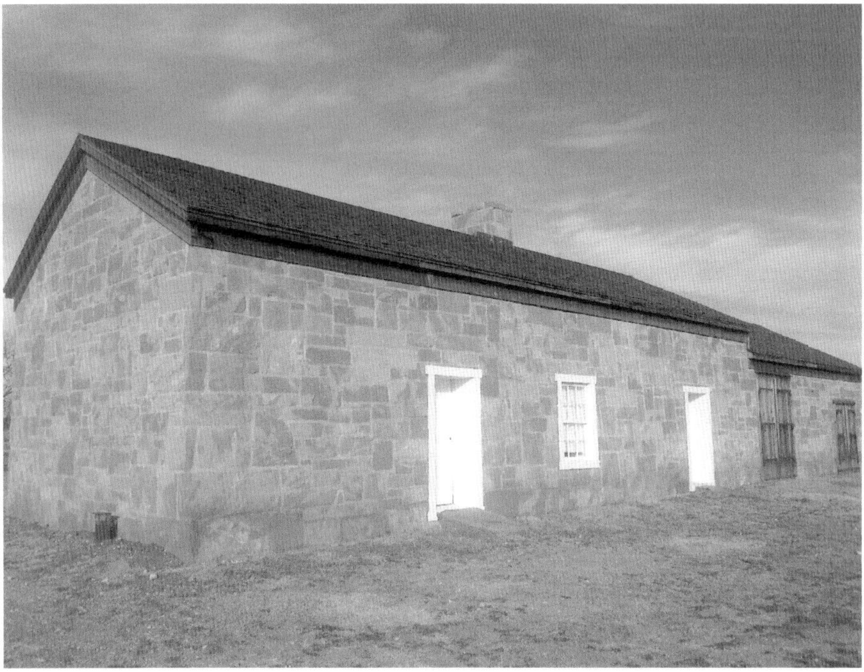

This reconstructed Butterfield stage stop at Fort Chadbourne, Texas, is now listed on the National Register of Historic Places. *Wikipedia.*

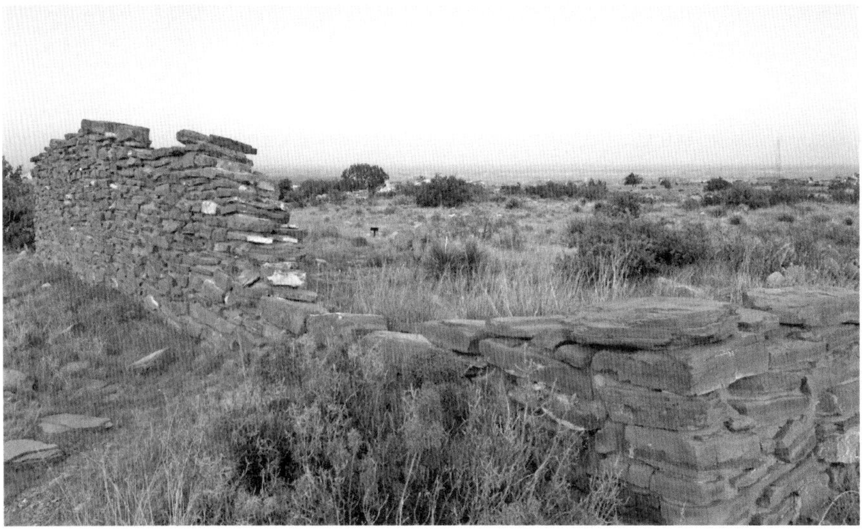

The Pinery Stage Station in Texas ceased operation in 1859 and is one of only a few company-built Butterfield stations left in the United States. *Photo by Rennett Stowe.*

horse wranglers, armed guards and cooks lived. In addition to a large room with one or two dining tables, it included crews' quarters for the drivers and conductors out on the road. Employees' families lived there, adding to the "homey" atmosphere.

The home station was pretty much a tiny town in itself, its complement of personnel often outnumbering those of a small fort. These stations grew into community hubs where ranchers and even those "in the area" would congregate for news, food and an occasional dance. People from as far as fifty miles away would come into "town" for socializing.

With the furnishings so pleasant and comfortable, passengers could opt to lay over at these home stations, resting up to continue the trip later. Meals consisting of bacon, beans, onions, coffee and meat were also provided at the home stations. Mention has been made of a concoction called "slum gullion," in which at least some of the ingredients were questionable. An unnamed through passenger wrote, "The fare, though rough, is better than could be expected so far from civilization and consists of bread, tea and fried steaks of bacon, venison, antelope or mule flesh (the latter tough enough), fried in a grease-encrusted and blackened skillet whose wielder seldom recognized the virtues of soap and water."

Another traveler said the fare was "…doughnuts green and poisonous with salterates, suspicious eggs in a massive greasy fritter, and rusty bacon intolerably fat." However, later on, another person wrote, "Our meals at the stage stations continued very good throughout the ride. The staples were bacon, eggs, hot biscuits, green tea and coffee, dried peaches and apples and pies that were uniform. Each meal was the same. Breakfast, dinner and supper were undistinguishable except by the hour."

Guess there's just no pleasing everybody. Meals cost between seventy-five cents and one dollar.

Activities around the home station were pretty much full time, as there were new horses and mules to be broke to harness, stages to be repaired, supplies to be maintained and food to be cooked. Stages rumbled through at least twice a week.

The swing stations, on the other hand, were set up with a minimum of facilities. The primary purpose was a place to change teams. Changing the horses or mules, a three- or four-man duty, involved unharnessing old teams and then harnessing new ones, as well as supplying rested animals with hay and water. The station included corrals for holding the animals and feeding and watering them. Some of these stations were little more than camps and were quite similar to the later "line camps" of the cowboy era.

The routine was to change horses or teams, make minor repairs to equipment and harnesses and then get back on the road. Passengers were urged not to dally and to "wash their faces" quickly. The personnel's only company was usually a quick "howdy" from the driver and passengers. Those working at the swing station were all skilled, well-armed men, and many of them enjoyed the company of Indians. More than a few times, Eastern passengers would stare wide-eyed at an Indian helping to change a wheel.

All stations were built with utility as a first consideration. Comfort was not one of their strong points, but near the far west and eastern terminus, habitation tended to soften the hard edges.

Generally, building the stations was labor intensive but not dangerous. However, Silas St. John's experience was out of the norm. He was in charge of building a line of stations from Stein's to Dragoon Springs in present-day eastern Arizona. The crew he hired to build the station at Dragoon Springs was a mixture of Americans and Mexicans. Before the arrival of the first stage, the Mexicans, using the cover of darkness, decided to take all equipment of any value and kill off the Americans. Silas was the only one to escape alive despite, according to several accounts, having his arm nearly chopped off. Nine days later, a stage from the west found him and patched him up long enough to get him to a doctor. He remained alive and employed by Overland Mail. He lived a long life without the use of one arm.

Stations were maintained at intervals of eight to twenty-five miles. At first, there were some runs of forty or fifty miles without a change of teams, but these were reduced until the average span was between ten and fifteen miles. Supplies of hay, grain and even water often had to be hauled long distances to the stations.

Another type of station, the terminus, was established at the division points, where passengers and mail could connect with other stage lines. Terminus stations were at Fort Smith, Arkansas; Fayetteville, Arkansas; La Mesilla, New Mexico; Yuma (Arizona City), Arizona; and Visalia, California.

Fort Smith was a place of personal business for John Butterfield. Here he maintained a large agency of the company, and due to the city's importance as a major division point, stables were regularly stocked with fifty to one hundred extra horses and mules. There was also space for five to eight additional stagecoaches.

Another important stop, the Boston Mountain range, was a rugged feature of the Overland route but had been well settled. Fayetteville was another terminus. In fact, Butterfield had his eldest son, John Jr., assigned as station agent there. Butterfield himself had a hotel and a large barn built and bought 360 acres of farmland on what would later become the outskirts of the western edge of the city. Today, the same route is known as the Boston Mountain Drive. Mr. Hiram Rumfield, a Butterfield employee who rode over the route, wrote the following to his wife: "It winds along the sides of the mountains over a surface covered with masses of broken rock and frequently runs in fearful proximity to the precipitous ravines of unknown depth."

La Mesilla, possibly the largest terminus, boasted an estimated population of 3,000 in 1858 (2,180 in the 2000 census). It was the largest settlement between St. Louis and San Francisco. Because of its location in southern New Mexico, it grew into a hub of transportation where other stage lines would intersect with Butterfield's. The Santa Fe Mail Branch Line and the San Antonio–San Diego Mail Line connected there, as did the irregular freight and passenger service Catlett Express.

Sam Bean, owner of the Overland Company's building and brother to Judge Roy Bean, said this about the first arrival of the stages: "For several days the community had sent riders out to locate and report back on the progress of the stages. The stage from the west arrived first, preceded by a charging rider. The news aroused the village, and soon the plaza was filled with a raucous crowd to greet this first stage."

Today's La Posta Restaurant is generally considered to be the debarkation point for Butterfield's passengers to lay over or connect with other stage lines. The divisional headquarters was a block west of La Posta in the building now housing El Patio Restaurant and Bar, directly south of the plaza. While El Patio faces the plaza, behind it was the Overland Mail Company's office and drivers' sleeping quarters. Albert Fountain II described the area as a large adobe building with a high-walled corral attached south of his new store and post office.

Arizona City (present-day Yuma) owes its existence to being a terminus for Butterfield's Overland Mail. Dutch emigrant Louis J. Jaeger ran a ferry service across the Colorado River beginning around 1851. Figuring that people would be flocking to the California gold fields, he charged two dollars for each trip across the river. During the next six years, he made between thirty thousand and forty thousand crossings, amassing a tidy sum. However, there was not a single house in the area except his brush lean-to. Across the river, however, was Fort Yuma, sporadically occupied by the U.S. Army.

A July 1854 surveying party looking for a transcontinental route approached Jaeger for service across the river. He quoted the small group

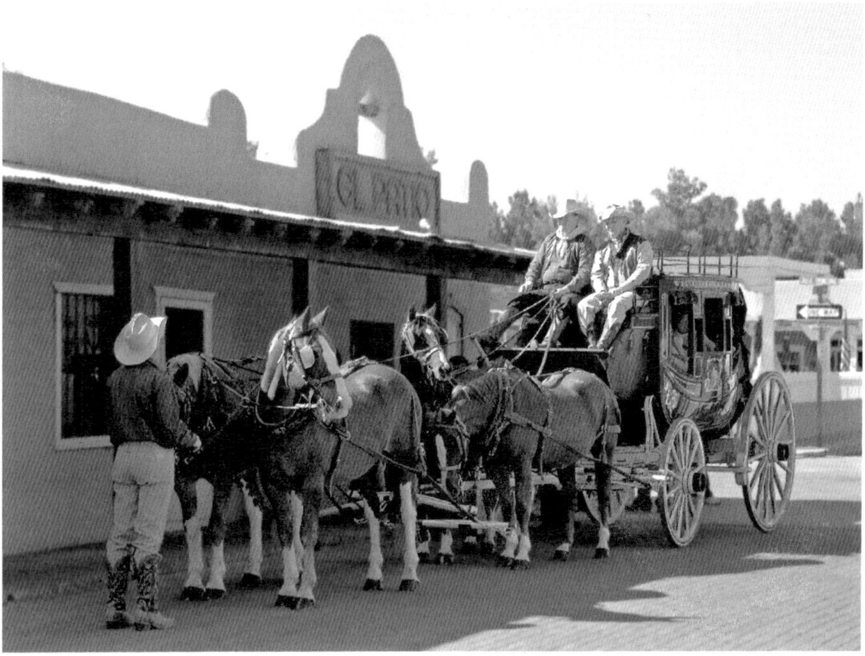

A stagecoach waits in front of El Patio Bar, originally owned by Sam and Roy Bean. The Butterfield headquarters were located directly behind this building on the same block. *Photo by Myke Groves.*

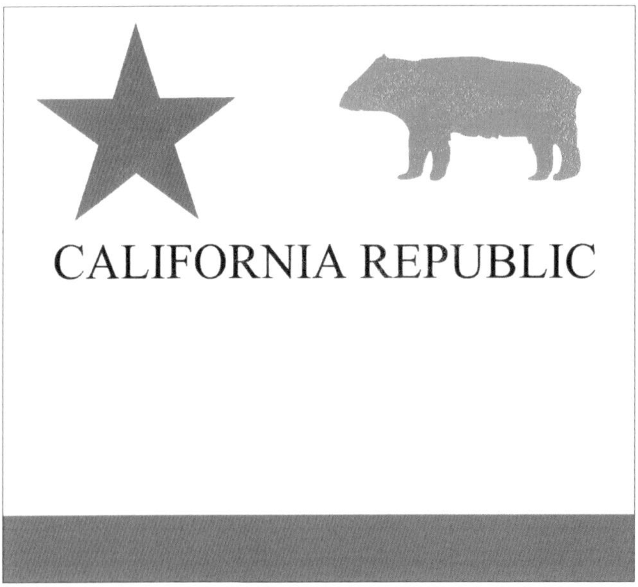

This original Bear Flag of the California Republic, designed by William L. Todd, nephew of Mary Todd Lincoln, was destroyed by fire following the 1906 San Francisco earthquake. *Public domain.*

fifty dollars, a substantial amount of money. The surveyors balked. So, in response, the surveying leader ordered his men to start laying out a town. Within an hour, the men marked off streets, laying out stakes for First Street and Central Avenue at the foot of Jaeger's ferry. "I like the looks of this town site," Jaeger is reported to have said, and he proceeded to buy a lot on the corner of Main and Central—for fifty dollars.

The surveyors pocketed the money, folded their maps and then handed fifty dollars to Jaeger for a ride across the Colorado. To his credit, the lead surveyor, who had no authority whatsoever to plot a town, did indeed record Jaeger's lot when they arrived in San Diego. Arizona City became a registered site, its sole inhabitant Louis Jaeger until 1856.

By 1858, when the Overland Mail Company began operations through that area, Arizona City had twelve souls. Within a year, that number had jumped to two hundred, and the city was later considered as a candidate for territory capital. The area was referred to as "that Yuma town," alluding to the Indian tribe in the area. Later, Arizona City became Yuma where Jaeger's original lot can still be found downtown.

Visalia, California, another terminus about 550 miles north of Yuma, was settled in 1851 by two brothers, Nathaniel and Abner Vise. Originally called Buena Vista, it grew from an outpost of two or three buildings to an actual settlement. The name was changed to reflect that of its founders. Most of the countryside was somewhat inhabited, and the trail was well defined. In 1852 Nathaniel wrote, "The town contains from 60–80 inhabitants, 30 of whom are children of school age. The town is located upon one of the subdivisions of the Kaweah [River] and is destined to be the county seat of Tulare." Early Visalia history indicates that a school and a Methodist Church were established the same year and that the following year, a gristmill and a general store were built. Visalia has been called a one-time "capital" of the buckaroos or vaqueros, the California cowboys.

In 1858, Visalia was added to John Butterfield's Overland Stage route as an important stop. Today, a plaque commemorating the location can be found at 116 East Main Street. Included in the early crop of citizens were some notorious individuals who preyed on the Butterfield Overland Mail and its passengers. Many saloons and hotels sprouted up around the stage stop downtown, and commerce was brisk, if not a bit risky.

Chapter 15

YOU'RE NOT GOING ANYWHERE WITHOUT WATER

Literally nothing could be accomplished in this venture without water. The question of where to find it and how to get it was paramount. Availability of water made the difference.

A few of the wells, such as the one at the Cornudas, New Mexico station, were hand dug by Indians. This water source, called Thorne's Well, was not a well but a cistern sheltered by a cave. In that area, early Indians left numerous petroglyphs (rock carvings) and pictographs (pictures painted on rocks) to watch over the water source. Carbon dating of the pictographs reveals the paint to be three hundred to four hundred years old.

Across the Sonoran Desert in Arizona, water was available at the various rivers and creeks, but when the route crossed from Maricopa Wells through a forty-mile stretch of desert, before striking the Gila River, the pickings were lean. From Fort Yuma to Carrizo Creek, California, about one hundred miles, the route was heavy with sand, and there was no water during the dry season.

When searching for a viable route, the lack of water was certainly taken into account. In a report for the Mexican Boundary Commission, John R. Bartlett reported, "Wells might easily be sunk so as to procure water at all times, in their various reconnaissances, discovered in many places not marked on the maps." In some instances, digging a shallow well would produce good water. The water table and underground springs and lakes were common, but many people were unaware of them. So when Butterfield's men began setting up stations along the route, they found potable water relatively close to the surface and were able to dig their wells.

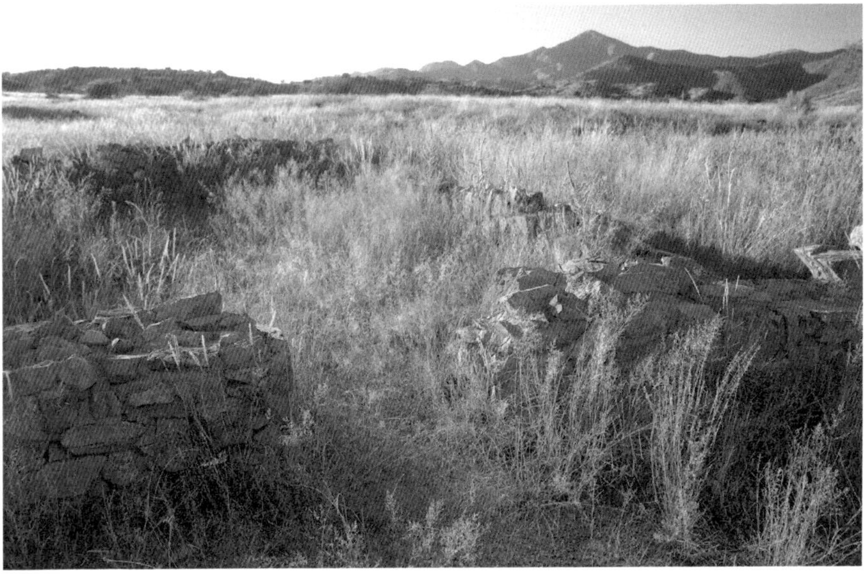

Ruins of the Apache Pass station in Arizona. *Photo by Myke Groves.*

However, thousands of dollars were spent on digging wells, damming up creek beds and building *acequias* (big ditches), as well as hauling water long distances. The original contract allowed for deviations along the route to ensure better access to water, which allowed for better traveling.

Availability of water was the reason Butterfield drew his route directly past Apache Springs (Puerto del Dado) in Arizona. This water was quite important not only to the station employees but also to the Chiricahua Apaches who lived there.

When Overland Mail representatives met with Cochise, head of the Chiricahua Apaches, the meeting was peaceful. Cochise, who had a hatred of Mexicans, looked at the Americans more as allies and not as enemies. He agreed to let them share the spring, his only stipulation being that the station be built a few hundred yards away from the spring. He wanted his people to be able to get water and not be exposed to direct contact with the Americans.

Butterfield's policies provided for common sense such as this. As a result, Cochise and the Chiricahuas remained on good terms with the Overland Mail Company. This system worked well for a couple years until an American lieutenant, recently out of West Point, combined with the shenanigans of a nefarious rustler changed the friendly feelings.

Chapter 16

STAGECOACHES: CRADLES ON WHEELS

What sort of conveyance would best transport the mail and passengers across this new transcontinental route? It wasn't until the early 1700s that wheeled vehicles came into general use in America. Famous Conestoga-type wagons appeared after 1750, and those big, awkward wagons carried supplies and people for the next one hundred years. At the same time, stagecoach service began, but that coach style was not suited for Butterfield's needs. John Butterfield knew his 2,800-mile route required specialized forms of transportation. After much thought and research, he decided the stately Concord stagecoach and Celerity wagons would do nicely—with some specific alterations.

Chapter 17
CONCORD STAGECOACHES: THE LUXURY LINERS

The Abbott-Downing Company of Concord, New Hampshire, manufactured the Concord coach. Founded in 1813 by J. Stephens Abbott and Lewis Downing, the company supplied coaches for all the important stage lines, both in the East and along the coasts of California and Oregon.

Considered the "luxury liner" of the day and costing $1,400 in 1857 dollars (the equivalent of $37,837 today), the two-and-one-quarter-ton Concord, much like today's Hummer, had room for six passengers inside, three to a seat. Often, many more rode on top. The front seat, under the driver's box, faced backward, while the rear seat faced frontward. Using the best materials available, it was built of stout white oak, ash and basswood and braced with iron bands.

The coach's body was suspended on two thorough braces, long leather straps extending the length of the coach. These braces were placed on top of one another to a thickness of about three inches. Used in the absence of steel springs to absorb the jars, this leather swing permitted the coach to rock forward and backward. While this new invention cushioned the ride, it also caused severe cases of motion sickness. Usually, after a day or two, the victim would recover and begin to enjoy his trip.

The most completely tanned leather and "iron of the best Norway stock" was used, according to a company flier. Each coach was a specialty item in itself, produced by a team of three or four men working together from beginning to end. Downing himself inspected each part of the Concord coach

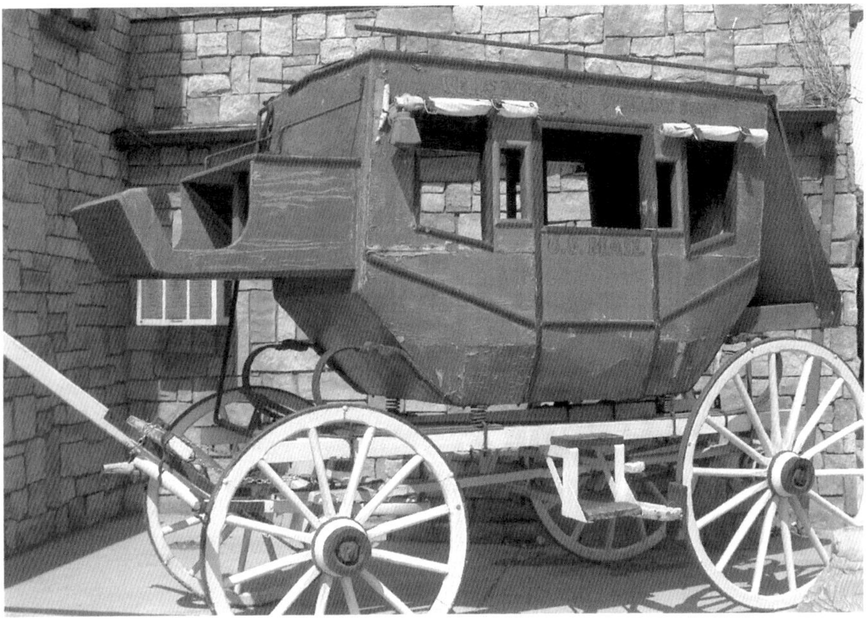

With a reputation for dependability and reliability, Wells Fargo expanded rapidly in the years 1855–66. The company is still around today. *Public domain.*

as it was being made. The story goes that if he found a part that wasn't up to his high stands, he ripped it off the assemblage and broke or smashed it front of the workman. It took three weeks to turn out one finished stagecoach.

Behind the body was the triangular "boot" used to carry mail or baggage. Up front, under the driver's box, was another boot or leather storage area, also used to carry excess. The mail was usually carried in the front and rear boots but could also be strapped on top or carried inside if necessary. Often passengers found themselves sitting with their feet on top of sacks of mail. Many men chose to ride up top, as the Concord's roof was solid.

The Concord stage was quite tall, the top being six feet, eight inches above the ground, not unlike some of today's motor coaches. Their doors had small glass windows, cutting down a bit on dust while also allowing in some light.

Where the roads were at least fairly good, Butterfield provided the Concord stages. However, this worked out to be only at the beginning of the route from Tipton, Missouri, to Fort Smith, Arkansas, and at the end of the line from San Jose to Yerba Buena (San Francisco), California. For the great majority of the time, the lighter, tougher Celerity coach was used.

Chapter 18

CELERITY STAGECOACHES: WHO ARE YOU CALLING A "MUD WAGON"?

The Celerity, built in Troy, New York, was designed for rough roads, thus making it one of the more popular wagons in the Southwest. It had canvas curtains that one could pull down over the windows and doors, which let in little to no light during a dusty trip. It was a true stagecoach built for the specific job of cross-country travel under rough conditions.

This wagon had a third bench seat in the center with a drop-down back. The center section of this middle seat was bolted to the floor, and the entire seat was located slightly off center to the rear, allowing more room for those sitting knee to knee. The two side sections were hinged so that they could be lifted for passengers to have access to the rear seat from either door. The backrest for the center seat was sort of a broad-banded harness that was suspended by straps from the inside roof. If the coach was not fully loaded, passengers could let down the backrest, leave the other seats in place and take turns "sleeping."

Since the stage ran around the clock, passengers tried to sleep on the center seat folded down. If arranged appropriately, up to ten men could sleep at the same time. The stage averaged 120 miles in twenty-four hours.

As a single passenger on the inaugural trip, Waterman L. Ormsby reported that he was able to raise the back out of the way, stretch out between the seats and sleep. Most passengers weren't that lucky. Complaints to management were loud and long. Passengers on the end of the benches had to hang an arm and a leg out the window. Of course, this meant that the canvas covering the window could not be lowered, allowing in more dust.

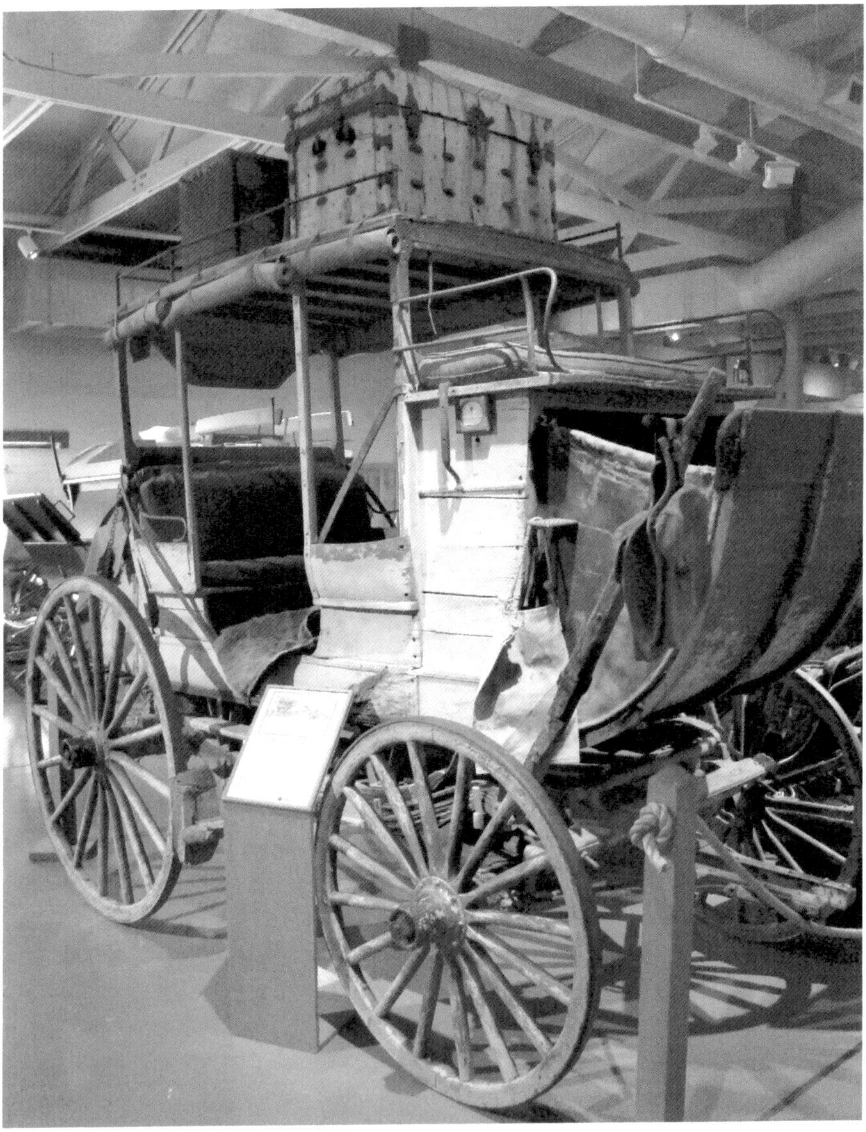

The trunk on the roof wasn't standard issue for Concord or Celerity stages. Instead, it was added by the coach operator. *Photo by Leon Worden.*

Unlike the Concord, the Celerity, which cost $500 ($13,500 today), had a canvas top, which could not support passengers or luggage.

Raphael Pumpelly, who traveled on the Celerity west to Tucson, recounted his ride: "…and there being room inside for only ten of the twelve legs, each side of the coach was graced by a foot, now dangling near the wheel, now trying in vain to find a place of support...Unusually heavy mail in the boot, by weighing down the rear, kept those of us who were on the front seat constantly bent forward."

Like the Concord, the Celerity was sprung on leather thorough braces, giving passengers an easier ride and mules and horses reduced strain. Due to the light canvas roof and lower supports of the thorough braces, the Celerity was a more stable coach for the rough Western roads. Where the trail was not well established, the Celerity could plow through sand and mud much more easily than the heavier Concord. Because of this, the Celerity stagecoach became known as a "mud wagon."

The best seat, all experienced passengers knew, was the one directly under the driver's box or bench. That meant the passenger would face backward, but the rocking wasn't as pronounced and the wind didn't hit him in the face. Often the veteran would give up his seat in the back to a newcomer and enjoy watching the "newbie" succumb to the coach's rocking.

Traveling thousands of miles over unknown territory through sand, rocks, mud and water, with no other help and survival of the people concerned at stake, dependability was no luxury—it was an absolute necessity. A simple thing like a wheel does not seem, at first blush, to have much to do with transcontinental travel, but without a proper wheel, none of this could have happened. It took someone knowledgeable and trained to get a stagecoach or other wheeled vehicle across the country, and when it broke down, finding a near "service station" was not possible. True, Butterfield established stage stations along the way, but the coach had to get there first. So, with all this in mind, Butterfield ordered a change of the wheel design after one or two trips.

It takes about half as much power to pull a four-inch-wide tire as it does a two-and-a-half-inch tire, so Butterfield requested that his tires be upgraded to this size. This was one of Butterfield's many innovations. Not only were the tires wider, allowing for better traction in sand and mud, but the wheels themselves were also larger and set farther apart than those found on the eastern stages. Set at sixty-two inches in track, they proved to be more stable. (As an aside, later on, Pontiac set the wheels of its automobiles at sixty-two inches. This was advertised to give better balance and lessen the danger of the vehicle turning over.) All the wheels were also "dished," meaning that the spokes flared outward from the hubs, allowed the weight to be carried by the lower part of the wheel. This had the effect of strengthening the wheel.

What the Concord and Celerity did have in common was another Butterfield first: bright paint. The coach body was painted a bright yellow,

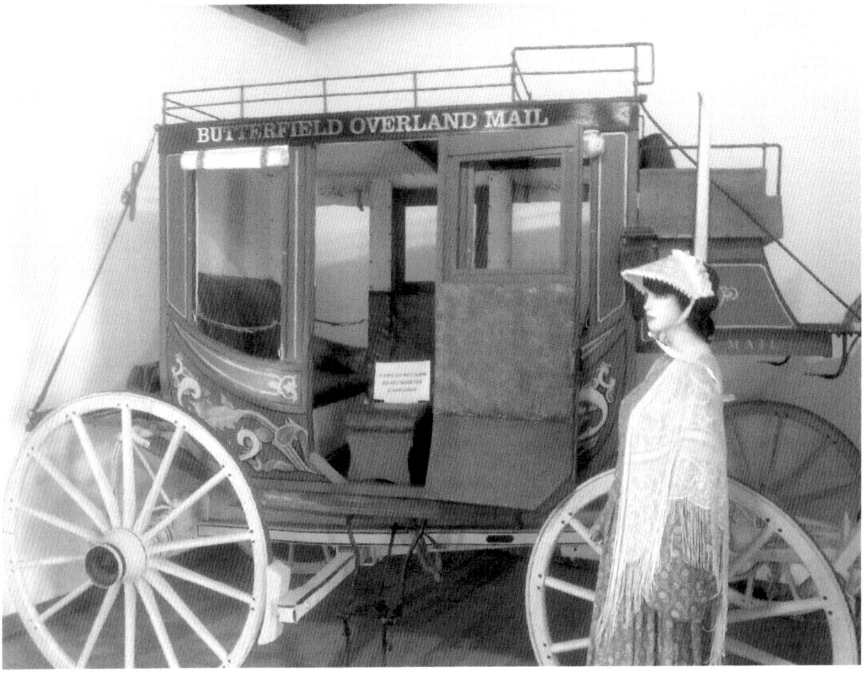

A Butterfield stagecoach displayed at Fort Chadbourne, Texas, which was established in 1852 to protect the western frontier. *Wikipedia.*

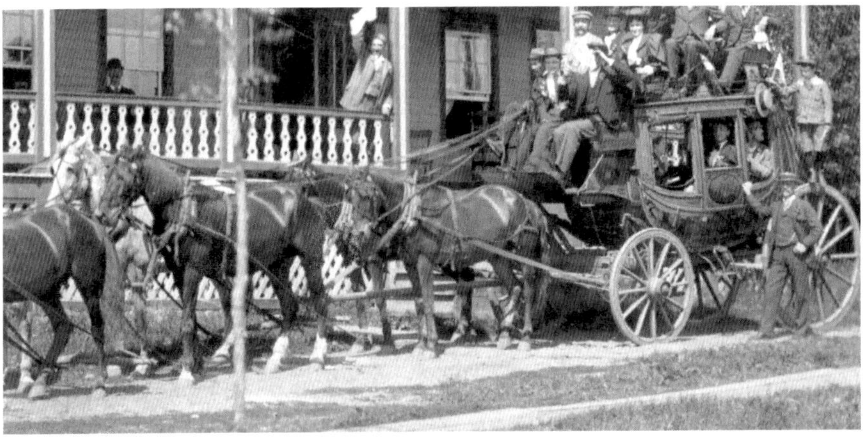

Friendly rivalries existed between the owners of early New York hotels. This Ampersand Hotel stagecoach was probably driven by Tuffield Latour. *Public domain.*

red, green or blue and marked by large letters denoting "OVERLAND MAIL COMPANY." The wheels and undercarriage were all painted a bright orange. There was absolutely no way to miss a Butterfield coach; even a coating of dust or mud couldn't hide the distinctive colors. Clean and shiny, they attracted attention. Butterfield wanted his coaches to stand out and be easily spotted in the vast desert. They did.

But stagecoaching wasn't all fun and games, as modern Hollywood would have us believe. Indeed, the travel was difficult and tested even the gentlest of humans. Passengers in the rear stared at the back of heads in the middle seat, and those riders had their knees dovetailed with the people facing backward. They rode with their baggage on their laps when it didn't fit in the boot or on top and occasionally had mail pouches at their feet.

The stagecoach traveled day and night, stopping ten minutes or less at the swing stations. At the home stations, passengers were allowed forty minutes for a meal. One passenger drew the following picture: "A through ticket and fifteen inches of seat, with a fat man on one side, a poor widow on the other, a baby on your lap, a bandbox over your head, and three or four more persons immediately in front, leaning against your knee, makes the picture, as well as your sleeping place for the trip." Another happy traveler noted, "The fatigue of uninterrupted traveling by day and night in a crowded coach, and in the most uncomfortable positions, was beginning to tell seriously on all the passengers, and was producing in me a condition bordering on insanity."

Chapter 19
DRIVERS: THE TRUE KINGS OF THE ROAD

While it took literally hundreds of personnel to get this amazing venture up and running, the one person on whom the achievement solely depended was the driver, also known as the "jehu," the "whip," "Charlie" and a number of monikers not fit to print in this family-friendly book.

Butterfield recruited a core of employees from his eastern companies. He knew all too well that driving in the East on established roads with repair stations nearby was quite different from what they would encounter in the West. And it was with this in mind that he chose his drivers. He made sure they would provide his operation with a professional air—something he had to have in order to keep the contract with the government.

The trail was divided into nine divisions, each possessing a superintendent and a number of drivers and conductors, also referred to as "shotgun guards." The driver and guard or messenger took a coach to the limit of his run or division and then turned it over to the team arriving from the opposite direction. They then took the arrived coach and returned it to its home station. It was much like the rental car process of today.

A division team, naturally, became very familiar with its stretch of road, which was imperative since it was often crossed at night. The drivers were literally the kings of the road—after all, they were in charge of making sure passengers and the mail and any packages arrived safely. They also had to be able to help change wheels, harness obstinate teams and fight off Indians. All while staying on schedule.

Butterfield kept his employees and stations well armed as part of his strategy of keeping peace with the Indians. His policy was to ask questions, ask more questions and then shoot if absolutely necessary. Walker Colt and Remington weapons accounted for 70 percent of sidearms during this period. The most popular gun in the box under the driver's seat was the double-barreled, muzzle-loading, caplock shotgun—hence the term "riding shotgun." In fact, hired gun Wyatt Earp rode "shotgun" for Wells Fargo in Tombstone, Arizona.

Butterfield chose only the best of the best. He personally selected each driver and closely watched how he handled the reins or ribbons. Afraid to lose the "feel" of the reins, the driver often worked through the cold without gloves. A first-rate driver could handle three pairs of loosely hitched horses, sending the lead team into a turn while keeping the wheel team on course until exactly the right moment for it to turn as well. The skills and sensitivity of a good driver were critical to safe coaching.

The term "jehu" (yay-who) was common for the period, and its origins could come from the Old Testament: "...and the driving is like the driving of Jehu, son of Hanani, for he driveth furiously." Another origin is cited as the Spanish pronunciation of the word "Jesus" (hay-soos). A scattering of profanities in both English and Spanish could be recognized anywhere across the desert.

Whatever you called him, the driver was boss. To drive a Butterfield stagecoach was akin to flying the Concord supersonic jet of today. The jehu stood a cut above the rest. He often walked around town with his classic black whip, spurs jingling on his high leather boots, his soft flat-topped felt hat and a yellow duster over his regular clothes and a strut in his step. He was paid big bucks, but he also had big responsibilities.

And yes, a few women drove stagecoaches, too. Possibly the most famous was Charley (Charles/Charlene/Charlotte) Parkhurst, who was hired as a man to drive stages in California. He did so for thirty years. It wasn't until his death in 1879 that people discovered he was a she—and had given birth years earlier. To our knowledge, Butterfield did not employ female stage drivers. Some of the colorful names of his drivers (ones that can be printed) include Big Foot Wallace, Peg Leg Pete and Jepe Hitchings.

Chapter 20
HORSES AND MULES: GIDDY UP!

Good roads and stagecoaches were a necessity to move passengers and mail across the United States. But nothing would move without horsepower, one of the most vital parts of the operation. Like car engines today, the horses did the work of pulling the load. Also like cars, they required feed and maintenance. The stagecoach business involved living animals that had individual personalities and capabilities. And because of that, there was a clear need to arrange them into a bona fide team so they would pull together and achieve the horsepower necessary to get and keep the coach moving.

The typical American stagecoach team was composed of four or six horses hitched in tandem. Each member of the team had a job to do, a part to play, and personality made a big difference. Because of this, they could not be interchanged easily.

The six-horse team was composed of the "lead" pair, the "swing" pair in the middle and the "wheel" pair directly in front of the coach. For slowing or backing up, only the wheel pair could be used, and its harness was essentially different from that of the other pairs.

Most of the horses were mustangs, "wild as deer and as active as antelope," according to the November 5, 1858 *San Francisco Bulletin*. They were all shod and branded "O.M." for Overland Mail. Despite being spirited mustangs, the lead horses had to be agile, quick and able to move sideways easily. The swing horses had to be steady and yet responsive. The wheel horses had to be strong and durable, as well as accustomed to pulling to the rear on the command of the driver.

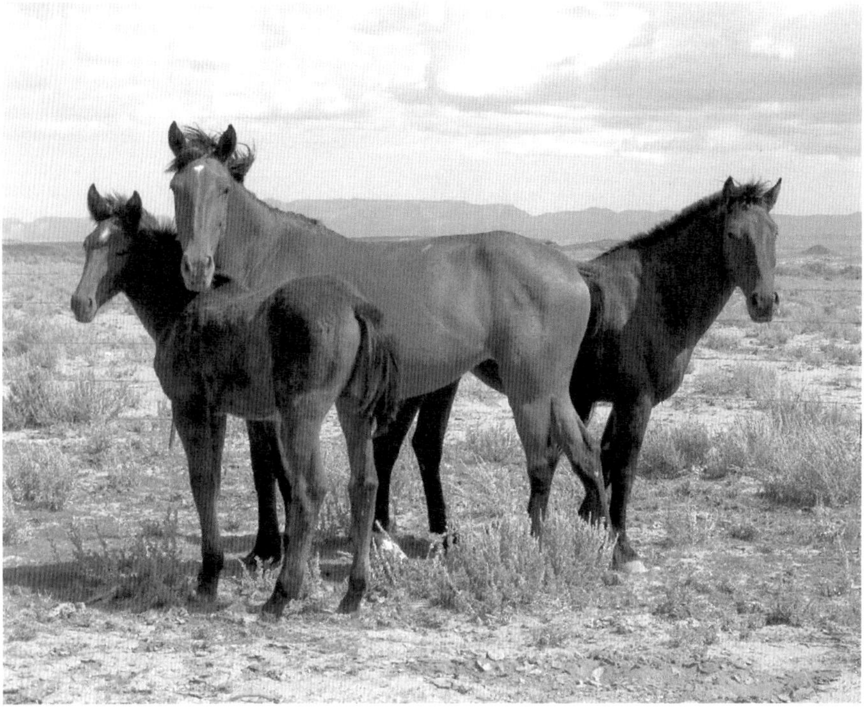

In 1971, Congress recognized mustangs as "living symbols of the historic and pioneer spirit of the West." *Image by John Harwood.*

In addition, there were differences between each of the paired horses. Regardless of pair, the horse on the left (from the driver's point of view) was called the "near" horse, while the one on the right was called the "off" horse. When another horse was needed for extra pulling power, it was always added in front of the lead team. This formation was commonly called a "wildcat" or "spike" team.

Butterfield always used the six-horse team configuration. Considering the weight of the wagons, along with that of passengers and mail, combined with the rough terrain, it makes sense to use such a team. When the driver would lace the reins in his fingers and get ready for the next part of his trip, he would get a feel for his team. During the first half hour of the trip, the jehu kept his eyes on his team. Even after the edge had been worn off, certain horses would react differently than others. The individual makeup of the horses varied from those that would bolt at the mere showing of a whip to those that would simply lay back their ears and sulk when popped on the rump. A six-horse team could be easily stampeded into a runaway in the hands of an inexperienced driver.

The animals relayed their current temperament, which an experienced driver would understand. The most significant signals came from the lead pair, as they alerted him to conditions. If one (or both) of the horses moved its head suddenly or more than normal, it was a signal to the driver. Something was up. More often, however, he was watching their ears. Even at dusk or dim light, the ears were usually visible, and the experienced driver watched the lead pair to signal what might lay ahead. This is what drivers meant when they spoke of "talking to the horses through the lines." A sulky or disconcerted animal would usually give itself away by having its ears fairly firmly laid back, showing that its attention was more on what was behind than what lay ahead. This was an indication that something was not going well. On the other hand, a relaxed and attentive animal would flex its ears forward and back, either one at a time or together, its focus on what was ahead.

The most common rate of speed was what is called a "hand canter" or "lope." This was the most durable pace with the most sustained continuity of effort. This slow gallop averaged about twelve miles per hour. Although this was the ideal pace, there were times—such as when a schedule needed to be met or when a jehu wanted to be a bit dramatic for passengers and observers when coming into town—when the horses galloped at a faster rate. A certain excitement was generated with the grinding of the brake and the dust billowing up under the wheels.

Butterfield passenger Raphael Pumpelly wrote the following about his trip around El Paso and through New Mexico:

> *Over the hard surface of this country, which is everywhere a natural road, we frequently traveled at great speed, with only half-broken teams. At several stations, six wild horses were hitched blind-folded into their places. When everything was ready, the blinds were removed at a signal from the driver, and the animals started off at a runaway speed, which they kept up without slackening, till the next station, generally twelve miles distant. In these cases the driver had no further control over his animals than the ability to guide them; to stop or even check them, was entirely beyond his power; the frightened horses fairly flying over the ground, and never stopping till they drew up exhausted at the next station.*

Stagecoach construction, harnesses and hitches were vital for Butterfield's success. The Overland Mail Company made innovations that became

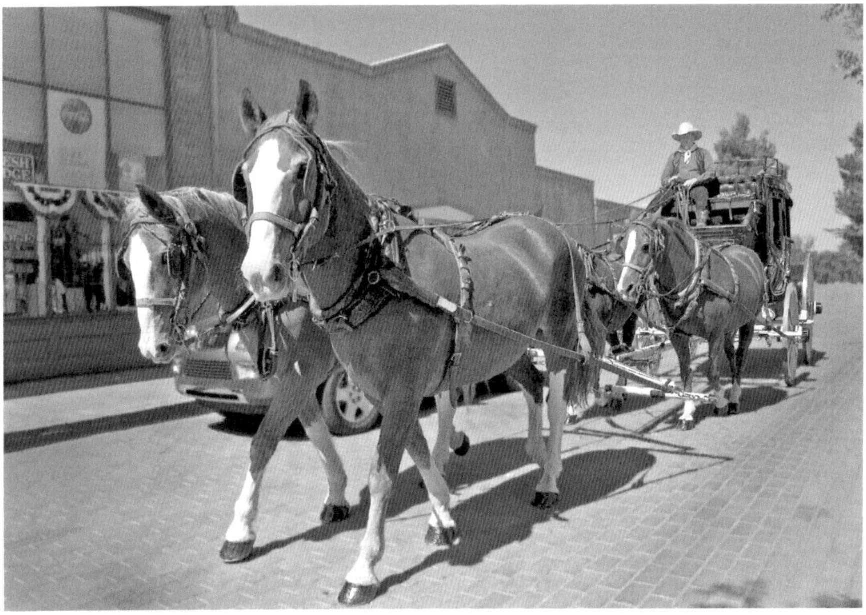

These horses show off for visitors in La Mesilla. *Photo by Myke Groves.*

standard throughout the West. All horses and mules had leather collars stuffed with excelsior or straw. This furnished padding between the harness and the animals. The collars were made with adjustable straps, but over a period of time, the collar became individually fitted to the shoulders of the animal. It became essential that each animal wear its own collar to avoid "galling" under a steady pull. A horse or mule with galls or big sores on its shoulders was unusable until healed.

While Hollywood has painted the image of powerful horses pulling these iconic stagecoaches, the truth is that mules were the primary pulling power across the deserts of Texas, New Mexico and Arizona. They had more stamina on less and rougher feed than horses or oxen. Butterfield brought in five hundred tough Missouri mules even though Spanish and Mexican mules were cheaper and readily available in the Southwest. Butterfield preferred the Missouri mules because of their size. However, he used horses on the better roads, where feed was plentiful and speed could be maximized.

While the mule was the powerhouse of the transcontinental stagecoach line, it did get a bit maligned. On the other hand (and possibly a bit glamorized) was the arrival of the stage in the southern New Mexico town of La Mesilla, population three thousand. The horse took center stage. Since it was the largest town between San Antonio and San Diego, La Mesilla quickly grew into a hub of transportation. Smaller coach lines used the town as a terminus, passengers connecting with Butterfield's Overland Mail line to head either west or east. Since the OM's scheduled service was only twice weekly, its arrival and departure were something of a social "do" around town. People came out to wave farewell or greet arrivals Butterfield insisted that since it was such a fanciful event, horses would be used to show off his brightly colored coaches. Once the fanfare had died down and the coach was well out of sight, the horses were traded for mules at the next station. The reverse was done when approaching La Mesilla. Usually, as long as the mules got fed, they didn't resent the fanfare given the horses.

While mules were desired to pull the heavy coaches, their meat was also desired by the Apaches. One big problem faced by the stage stations was that of keeping the mules safe. They installed armed guards and built sturdy corrals, but somehow the Apaches managed to take more than a few. This problem was felt mostly with other stage lines, as Butterfield had a good relationship with the Apaches. Indian raids were the main reason the San Diego–San Antonio Line closed down in 1861, even though it shared almost nine hundred miles of road and several stations with Butterfield's line.

PART IV
THE TRIP ITSELF

Chapter 21

OPENING DAY: WHAT A SEND-OFF!

"Remember, boys. Nothing on God's earth should stop the mail!" proclaimed John Butterfield. And nothing did.

Surprisingly, after all the work and hoopla surrounding the controversial enterprise of creating a transcontinental mail route, one would expect endless and boisterous celebrations lasting for days until the next stagecoach showed up. In reality, the send-off for one of the greatest national experiments to date roused little interest. The coaches were scheduled to leave San Francisco, California, and Tipton, Missouri, simultaneously on September 15, 1858. But they didn't. The eastbound left first, on September 15, while the westbound headed out on September 16. Except for a couple glitches, the operation started out smoothly.

Waterman L. Ormsby, a correspondent for the *New York Herald* and Butterfield's first through passenger wrote, "As the service is to commence on the Pacific side at the same date, we shall have, if the contract is fulfilled, a semi-weekly mail arriving in San Francisco, Memphis and St. Louis, completely eclipsing the news of the steamers."

The plan for the beginning of such an auspicious occasion was for John Butterfield himself to take the bags from the post office at St. Louis, carry them to the train station and deposit them in Tipton, Missouri, where they would be placed on the stagecoach.

But there's something about best-laid plans, and it turns out that the postmaster had already sent them ahead—rather unceremoniously—to the railroad depot in a regular mail wagon. So, the first portion didn't quite work out as planned. Undaunted, Butterfield charged ahead to the train station and took charge of them there. Everyone knew there was little margin for error, as the entire system went into effect at one time. On the first trip, the terms of the contract had to be fulfilled in order to consider it secure.

The first leg of the journey, 160 miles by train from St. Louis to Tipton, went smoothly. At Tipton, the end of the Pacific Railroad at that time, the mail was transferred from the train cars to the awaiting stagecoach. Ormsby, who was keeping a close track of the time and miles, wrote, "Thus far we are up to time."

When the train arrived at Tipton, everything was ready for the mail exchange, and John Butterfield, excited and impatient, paced and helped until his son John Jr. climbed aboard and picked up the reins. John Sr., known as "Old John"—because at fifty-seven, he was considered old—scrambled into the Concord stagecoach, ready for his jaunt to Fayetteville. John Jr. was awarded the opportunity to write history as he took the reins of the first transcontinental mail route.

The Butterfield Overland Company operation was definitely a family affair. Not only did John Jr. get to drive the first stage, but John Sr.'s third son, Daniel, figured out the schedule. Daniel later became known as the man who wrote the words for "Taps"—in honor of his father.

And so it began. The first westbound stage left Tipton, Missouri, at 6:00 p.m. on Thursday, September 16, 1858.

However, there were no gathering crowds, no cheers, no bands, no ribbon cutting, no parades, no ceremony. Nothing. But with a flick of the "ribbons," the first stage with six good horses was off on the first schedule west. One report states that one of the less-than-interested onlookers merely mumbled, "Goodbye, John" as he drove away. It wasn't clear to which "John" he spoke.

Things improved. The first acknowledgement of any consequence occurred at Fort Smith, Arkansas, where the stage rolled in at 2:00 a.m. Plans had called for them to meet the mail from Memphis, and they had expected to beat its arrival but found the stage from Memphis had actually beaten them by fifteen minutes. The news that both mails had arrived spread rapidly, and soon houses were lit up and horns blown as people gathered at the hotel to take a look at the first stages.

On the other coast, the eastbound stage out of San Francisco, which left on Friday, September 16, received only a slightly better send-off. The headquarters on Montgomery and Market saw a few moderately interested well-wishers but certainly not the crowds that were expected.

With the lukewarm send-off experienced by the Overland Company, three weeks later, when the stage made it into central California, the greeting of the first batch of mail from the East was an unexpected jaw-dropper.

People gathered all along the route waiting anxiously to catch sight of the stagecoach that had traversed the entire country.

Some of the loudest and most raucous fanfare occurred in Visalia, some 150 miles south of San Francisco. The stage got a series of "anvil" salutes that sounded more like canon shot and was a practice reserved for special occasions only. The salute was done by filling the square hole on the top of one anvil with gunpowder and then placing another anvil upside down on the top of the first. When this was set off, the top anvil flipped up in the air, and the loud report could be heard for miles. "The rousing cheers they gave us as we drove off… ought to be remembered in the history of the town, so I here immortalize them," Ormsby wrote.

With that type of welcoming fanfare in mind, when the stage finally arrived in San Francisco, the driver really "let the team out" and clattered through the streets. The brightly colored stage and galloping horses drew up at the post office in the plaza. The driver jumped off the box, hauled out the mailbags and rushed to the post office door. He yanked, but the door was locked. He shouted and banged on it until 7:30 a.m., when an irritated man unlocked the door. "You're early," he growled. Sunday, October 10, 1858, would live in infamy.

The biggest hoopla occurred in the East—but only upon the completion of the first trip. In Fort Smith, when the stage arrived on October 13, everything was in place. The town was decorated with flags and bunting. Business was suspended. John Butterfield led a parade, riding in one of his own coaches. Gaily decorated floats, a company of soldiers from Fort Smith and a brass band heralded a new era. Firemen paraded in red shirts and helmets while other dignitaries followed in five more coaches. Even a circus, which happened to be in town, joined the parade.

The party, filling a local park, continued with several hours of speeches, including one by Butterfield. A banquet and ball climaxed the festivities. Topping off a glorious acknowledgment of his endeavor, John Butterfield was carried into the banquet hall on the shoulders of four strapping young men.

When the first stagecoach mail eastbound was delivered in St. Louis, the *Daily Alta California* reported, "Cheering crowds, led by the St. Louis Silver Band, escorted the driver and passengers to the Planter's Hotel…Once there more toasts were drunk to the success of John Butterfield and his bridging of the continent."

Even President James Buchanan had something to say: "I cordially congratulate you upon the result. It is a glorious triumph for civilization and the Union."

Chapter 22

FROM SEA TO SEA: WHAT WAS IT LIKE, REALLY?

So, after the excitement died down, what were the day-to-day operations like? After all, 2,800 miles is a long way to go, and with all the personnel, animals and, of course, mail and passengers, things were bound to go wrong—but perhaps more importantly, they were bound to go right.

"He knows that it brings tidings from the hearts and homes he left behind him," an Overland passenger wrote of the mail in the *New York Post*. "It binds him stronger and firmer to his beloved country. So it's regular arrival that the inhabitants know almost the hour and the minute when the welcome sound of the post horn will reach them. The Overland is the most popular institution of the Far West."

The Overland Mail service grew in favor—so much so that by 1860, more letters were sent over the Butterfield route than by the ocean steamers. Even in England, sealed letter bags were made up regularly for San Francisco and the English Pacific Coast. The British even sent their mail to Canada by way of the Overland Mail.

The trip was not inexpensive. For through travelers (ones who went coast to coast), the fare was initially \$200 either way—close to \$5,714 in today's market. In 1859, it was changed to \$150 from San Francisco, while the westbound price remained at \$200. Later, the fare was reduced to \$100 in 1860. Eventually, after a whole lot of confusion and head scratching, a price of \$150 was chosen, regardless of direction.

The fare for "one way" passengers was set at \$0.10 (\$2.70 today) per mile for the distance traveled. Luggage was limited to forty pounds per

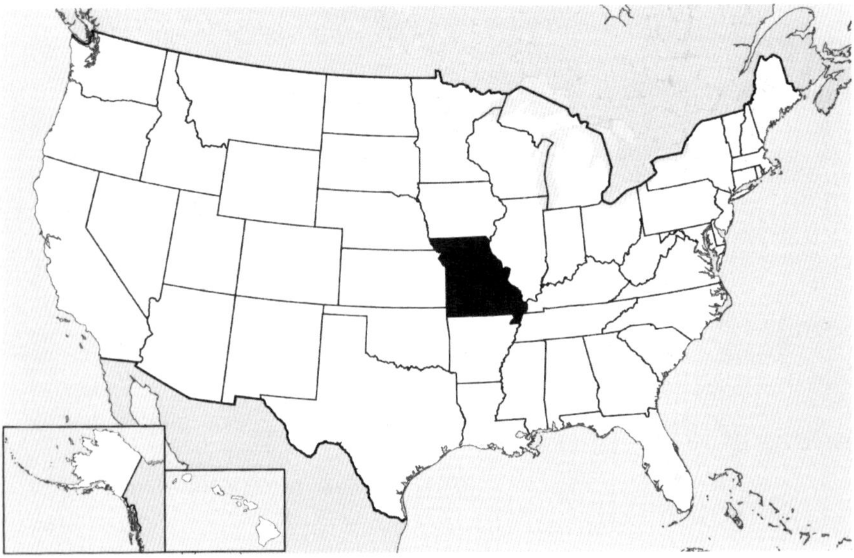

Missouri Territory was part of the Louisiana Purchase. Part of the territory was admitted as the twenty-fourth state on August 10, 1821. *Public domain.*

passenger, with an extra charge added on for more weight. Is this starting to sound familiar? Luggage in excess of forty pounds could be refused if space was limited.

Much can be written, said and speculated about the trip itself. One of the best accounts of the amazing journey was chronicled by the previously mentioned Waterman L. Ormsby. It is because of his letters home and to his newspaper that we have an almost mile-by-mile accounting of the trip. What a fantastic opportunity to see through his eyes this unfurling country!

To prepare for such a journey, a San Diego newspaper recommended that passengers bring along "one Sharp's rifle and a hundred cartridges; a Colts navy revolver and two pounds of balls; a knife and sheath; a pair of thick boots and woolen pants; a half dozen pairs of thick woolen socks; six undershirts; three woolen overshirts; a wide-awake hat; a cheap sack coat; a solder's overcoat; one pair of blankets in summer and two in winter; a piece of India rubber cloth for blankets; a pair of gauntlets; a small bag of needles and pins; a sponge, hair brush, comb, soap, etc. in an oil silk bag; two pairs of thick drawers; and three or four towels."

The Pacific Railroad left St. Louis, Missouri, as scheduled on 8:00 a.m., September 16, 1857. The tracks headed due west, followed the Missouri River to Jefferson City and then veered slightly north to the new village of

Tipton, thirty-eight miles farther west. Tipton, the western terminus for the railroad, had sprouted a mere six months before the Overland Mail met the railroad.

When the train reached Tipton, the bags were placed on the brand-new, custom-built Concord coaches with the Overland Mail Company logo emblazoned above the door. The passengers who boarded included not only Waterman L. Ormsby but also John Butterfield Sr., Judge Wheeler (not a real judge but a newspaper publisher) and his wife and their two children along with two other men. Their arrival in Tipton had been behind schedule, and in a hurry to make it up, not a minute was spared. In fact, it took only nine minutes from train arrival to stagecoach departure. "We found the first coach ready, the six horses all harnessed and hitched, and Mr. John Butterfield Jr. impatient to be off," reported Ormsby.

John Jr. drove the first leg of the new route and apparently was determined to stay on schedule. Ormsby said, "To gain thirty-two hours, as we have, on the already close timetable…has not given us much time to go easy over the stones." He gave up writing until later. "Yet the new scenes which constantly meet the view, the variegated aspect of the country, the curious characters to be met, and the novelty of roughing it overland, are, I think, quite a recompense for any slight inconvenience which may be experienced."

The first stop at Shackleford's, about seven miles from Tipton, went as planned. The coach braked in front of the home station, where food awaited the passengers. While they ate, the team was changed and harnessed. As soon as supper was over, the passengers once again boarded the stage. All of this was done in twenty minutes.

And so they continued on schedule. In fact, the stage, pulled by four horses, took only one hour and forty-five minutes to get thirteen miles—all uphill. That was considered pretty good time. Not all went according to plan, however. At one way station, John Jr. traded off with a driver who wasn't feeling well and didn't know the road as well as he should. They had to take it slow since it was dark and the coach lamps were not illuminating well. It took three hours to travel sixteen miles to Warsaw, Missouri, on the Osage River. Even so, they remained ahead of schedule. "As we neared the stations we blew our horns to apprise them of our coming," wrote Ormsby. He notes the time as 3:00 a.m. and that they found the fresh horses already harnessed. Within moments, they were back on the road.

Ormsby again draws us a picture: "As we left Warsaw, we had to be preceded by a man on horseback, with a light to show us the way through the ford [across the Osage], but the river being rather low there was not much

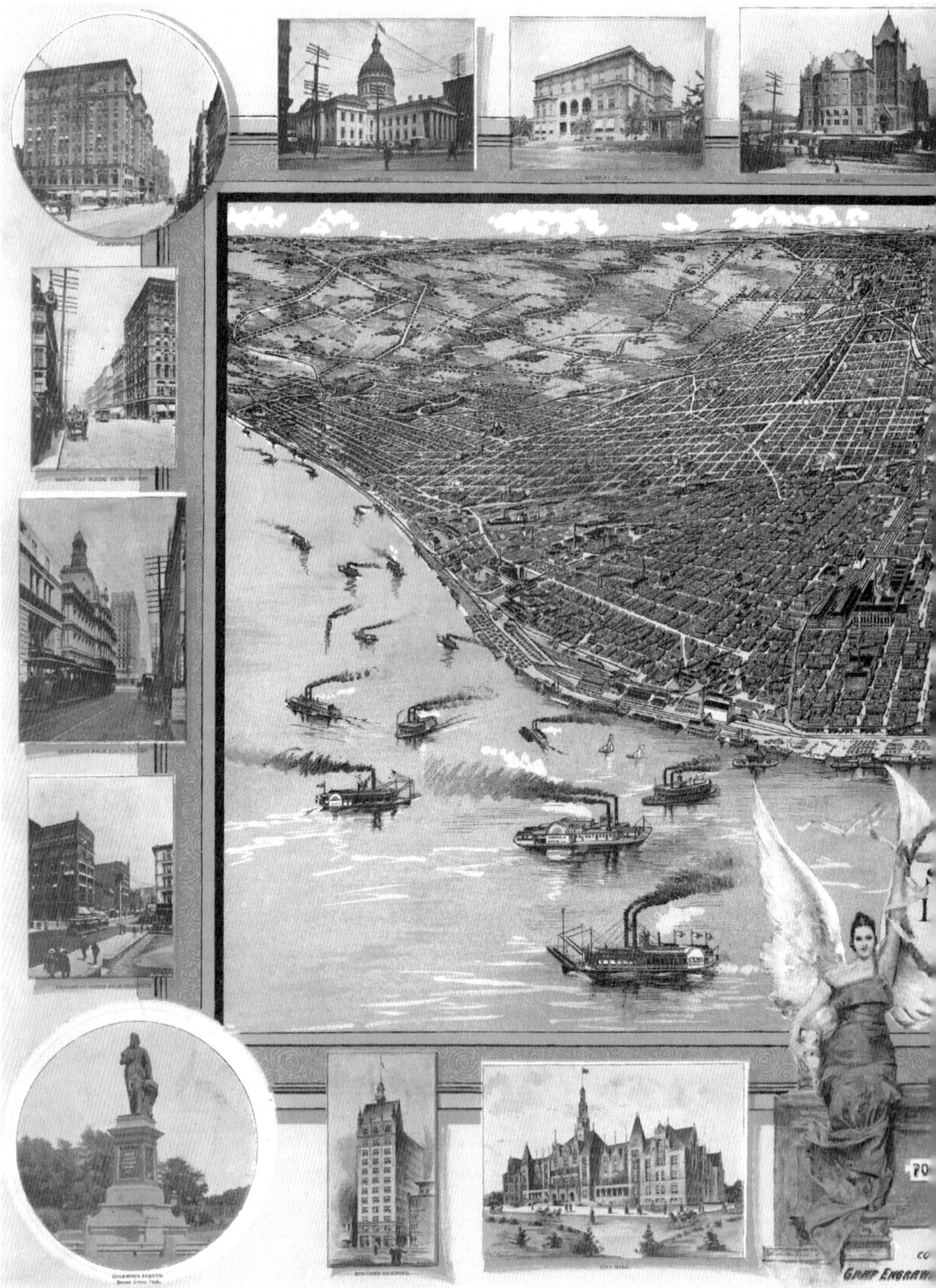

A birds-eye view of St. Louis, 1896. *Illustrated map by F. Graf.*

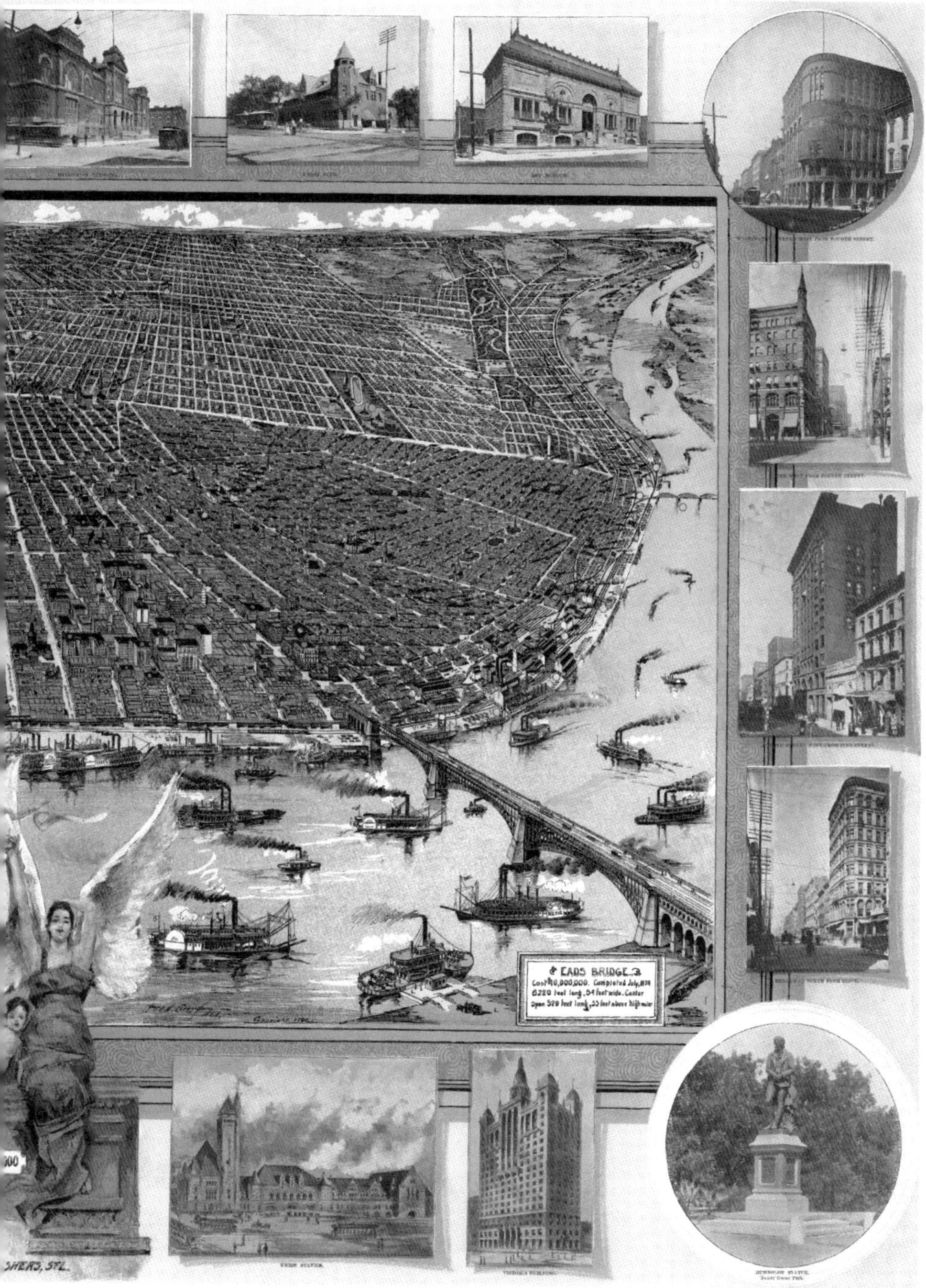
EADS BRIDGE
Cost $10,000,000. Completed July, 1874
6220 feet long, 54 feet wide. Center
Span 520 feet long, 55 feet above high water
UNION STATION
HUMBOLDT STATUE

danger…the music of the forest, the moonlight struggling through the trees, the easy motion of the vehicle moving on the sea, all tended to make one thoughtful of the impressiveness of the occasion." Indeed, it was impressive.

When they arrived in Springfield, a flourishing town of about two thousand, people gathered to see the first overland mail and to congratulate both Butterfields on their accomplishments. The fanfare ended with a salute of several guns. It was there in Springfield that they said goodbye to the large Concord coach and boarded the rugged, sturdier Celerity wagon. Ormsby found it "a very agreeable bed for one." One can only wonder what he'd say if there were nine others lying with him.

They left Springfield at 4:00 p.m., spending only forty-five minutes in town. After taking on small mailbags at the post office, they picked up a man whose only reason for riding was to be on the first overland coach out of Springfield. With a flick of the reins and a "Step up, now!" from the driver, they loped out of town.

They traveled all day and all night, stopping only to change horses and grab a bite to eat. Close to a day ahead of schedule, at breakfast time on Saturday, September 18, they arrived at Callahan's, a stage stop about twelve miles from Fayetteville, Arkansas, near the border of Indian Territory. Within a short amount of time, the wagon was greased, the horses changed and breakfast eaten. And then they were off again. "The route leads over those steep and rugged hills which surround the Ozark range in this section of Arkansas, and we were just three hours going from Callahan's to Fayetteville," Ormsby wrote in his report to the *Herald*. This portion of the route was twelve miles.

Fayetteville at that time boasted two churches, a county courthouse, various stores and close to two thousand citizens. Long before his Overland dream became reality, John Sr. had recognized the need for a fine hotel and planned to build one there. He also had bought several acres of property.

All was on schedule, with only a few minor difficulties having been thrown in their path. Shortly before noon, a fresh driver climbed up on the box seat, flicked the reins over the horses and again the wheels turned. At this point, they were twenty-two hours and seventeen minutes ahead of schedule. (Today, this route has been partially preserved as Arkansas Highway 265 and, in Fayetteville, Old Wire Road is designated with historical makers. In south Washington County, Arkansas, another historical segment exists along Bugscuffle Road.)

Before trying to navigate the Boston Mountains, they exchanged the horses for four strong mules. "I might say our road was steep, rugged, jagged, rough and mountainous—and then wish for some more expressive words," reported Ormsby. The road wound along mountain cliffs and ridges and through white oak stands. After several hours and nineteen miles of nail-biting braking and dragging and mighty efforts by the mules and driver, they arrived at the next station relatively unscathed. The mules were exchanged for horses for the trip into Fort Smith.

Before reaching that destination, they crossed the Arkansas River on a flatboat at the flourishing little town of Van Buren. Since it was nighttime, a mounted guide with a lantern helped them cross the flats and up the steep sandy bank. They arrived at Fort Smith at 2:05 a.m., having traversed the sixty-five miles from Fayetteville to Fort Smith in fourteen hours and seven minutes.

At that time, Fort Smith (established as an army post in 1817 and incorporated in 1842) was a thriving town of 2,500 and served as the hub of several shorter mail and passenger routes. Even so, upon the Overland's arrival, horns were blown and houses lit up, and many flocked to the hotel to have a look at the wagons and mail bags.

In comparison to the rest of the trip so far, an exorbitant amount of time was spent in Fort Smith—one hour and twenty-five minutes! In that time, they examined the various mails and sorted the waybills, which were stage line advertisements posted in towns and villages. They also joined the two mails from Memphis and St. Louis, changed stages and teams and dropped off passengers. This was where John Butterfield Sr. and his son left the adventure.

Orsmby wrote of Butterfield, "He is the most energetic president of a company I ever saw. He appears to know every foot of ground and to be known by everybody, while his son John has been very active in getting good stock on this end of the route…If the Overland Mail does not succeed, it will not be for lack of his arduous personal exertions. He urged the men in changing horses at every station, often taking hold to help, and on one occasion driving for a short distance."

At 3:30 p.m. on Sunday, September 19, the stage rolled out of Fort Smith exactly twenty-four hours ahead of schedule. They were now 468 miles out of St. Louis.

From Fort Smith west into Indian Territory (present-day Oklahoma), the Choctaws watched the stage rumble through their land. Ormsby reported seeing many quite wealthy Indians, preferring to raise cattle instead of farming. The land, he said, was well adapted for raising corn, tobacco and hemp. He waved at the various Indians as he passed by, but

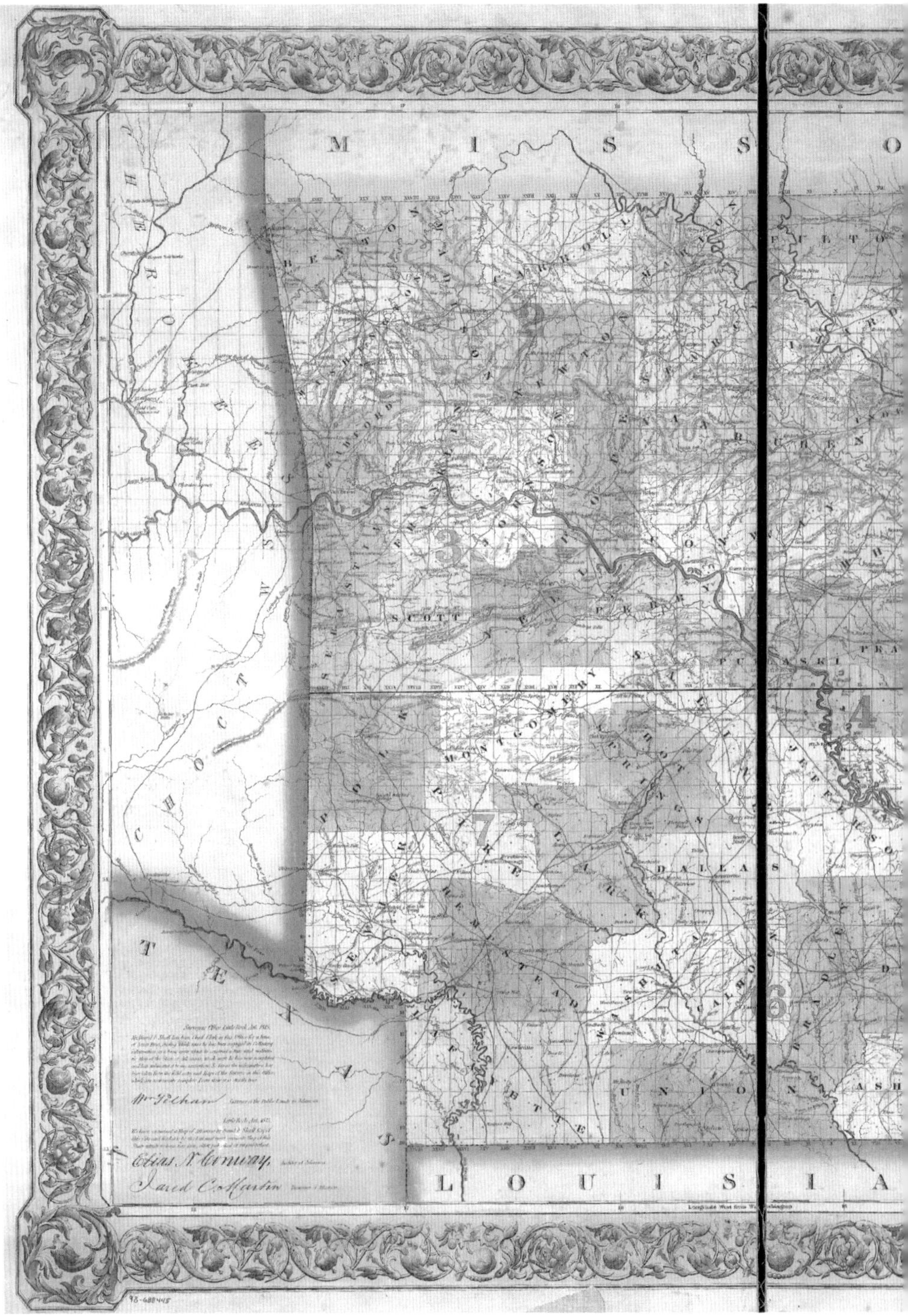
SCOTT
MONTGOMERY
PULASKI
DALLAS
UNION

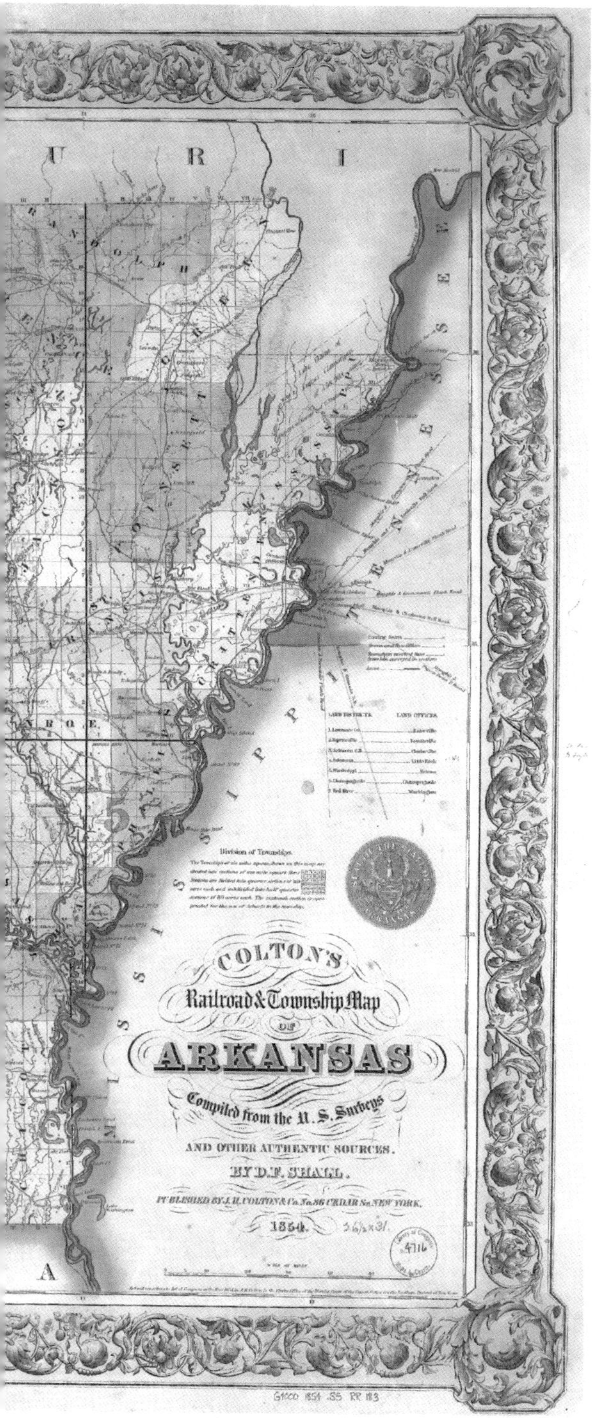

"Arkansas" is a French pronunciation of a Quapaw word meaning "land of downriver people." Admitted to the United States in 1836, Arkansas withdrew in 1861 to join the Confederacy. *Public domain.*

they "seldom said anything beyond 'Good Day'—and only when spoken to."

Seventeen miles past crossing the Arkansas River, the territory's governor, Tandy C. Walker, used his home as a station stop for the Overland Mail. This town, Skullyville, was established in 1832 as an Indian agency. The governor's house was the old agency house where government annuities were paid to Indian expatriates. On this special occasion of the initial Overland Mail meeting, Walker himself came out and helped hitch the horses.

Along the way, after leaving Skullyville, the stage passed many emigrants from Texas in covered wagons camping at night and traveling by day. The road thus far had been rocky, smooth, downhill, uphill, sandy and hazardous, but the stage made it through. About sixty miles north of

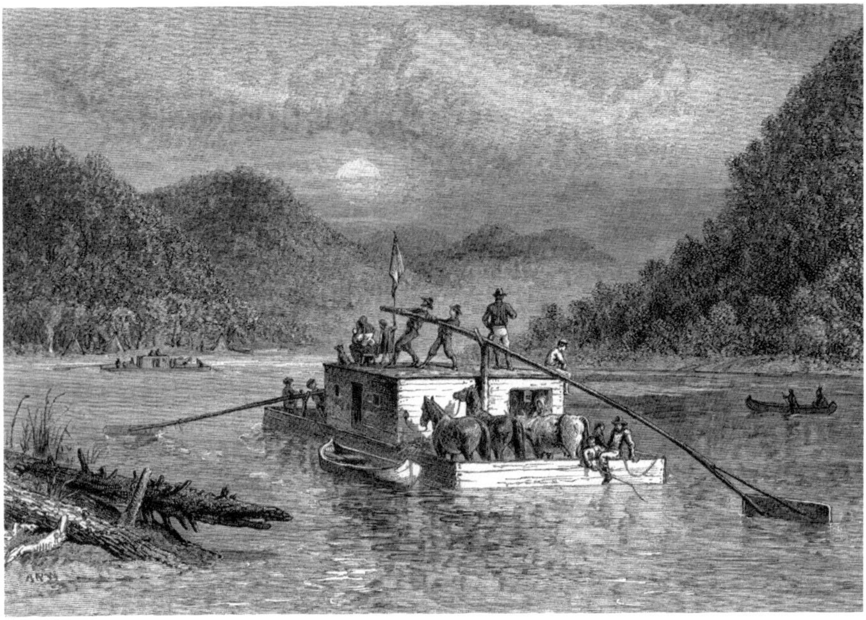

Rectangular, flat-bottomed boats without keels were easy to build. *Circa 1788 wood engraving by Alfred B. Waud.*

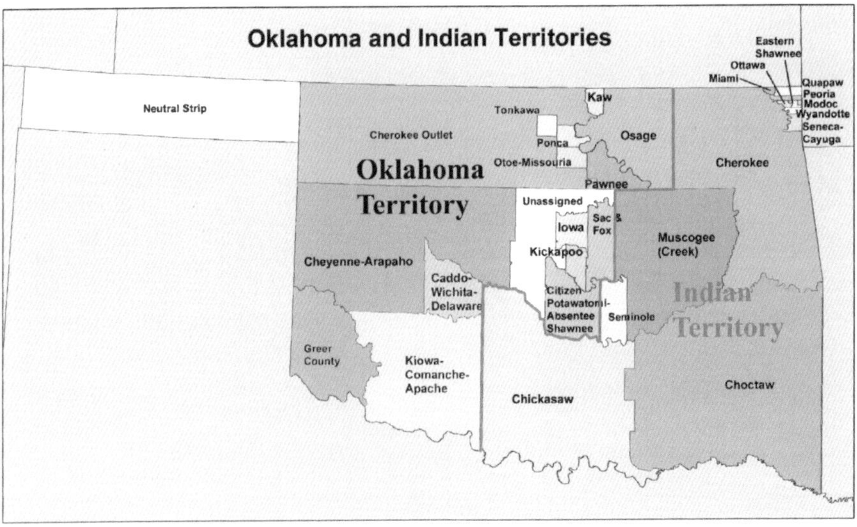

This map of Oklahoma was created using 1890 census information. *Public domain.*

the Red River, which defines the Texas border, they encountered a patch of woods that was "tortuous and stony." The driver's ambition to make good time, according to Ormsby, overcame his caution. "To feel oneself bouncing—now on the hard seat, now against the roof, and now against the side of the wagon—was no joke," he wrote. "I can truthfully say that I rather liked the excitement of the thing." However, shortly after those thoughts came two heavy thumps and a lurch of the wagon that unseated everyone. A crushing sound denoted that something had broken. Upon stopping, they discovered that the only casualties were a seat and a split wagon tongue. After repairing the damages at a way station, they made the next eighteen miles in two hours and fifteen minutes.

By the time they reached Colbert's Ferry on the Red River on Monday, the twentieth, at 9:50 a.m., they were thirty-four hours ahead of schedule. The problem with being so fast, however, was that now they had to wait for another team coming to relieve their team and to carry the mail. They spent the time sleeping, eating and swapping stories. Benjamin F. Colbert (1828–1893), of Chickasaw parents, had moved to the Red River area in the early '40s and became one of the wealthiest citizens in the Indian Nation. In 1858, the Chickasaw legislature granted him permission to build a ferry across the Red River. Married four times, he fathered fourteen children.

Whether ahead of schedule or right on it, it took eight days to cross Texas. Eight. And the miles were mainly across uninhabited low hills, gullies or dry creek beds, washes, sand dunes and mesquite. The only things to break the monotony were wind, rain and spotting an occasional fort. Thankfully for Butterfield, unlike many of the other passenger/mail services, Indian attacks were not an issue on this run or any subsequent runs.

From Colbert's Ferry, the stage arrived in Sherman, a town of about six hundred in northeastern Texas. As usual, the stationmaster had a fresh team ready to go. The stage departed at 4:40 p.m. on Monday, September 20—four days, six hours and twenty minutes from their departure time from St. Louis, a distance of 673 miles. At this point, they were almost thirty-two hours ahead of schedule. (Months later, in early 1859, Sherman was made a distribution point, through which Texas settlements were given postal service.)

Along the way, especially along the Texas plains, the home and way stations had not been completely finished. Ormsby noted of one of the stations, "Breakfast was served on the bottom of a candle box, and such as we were perched on inverted pails. There were no plates and but four tin cups for the six of us, including drivers and workmen."

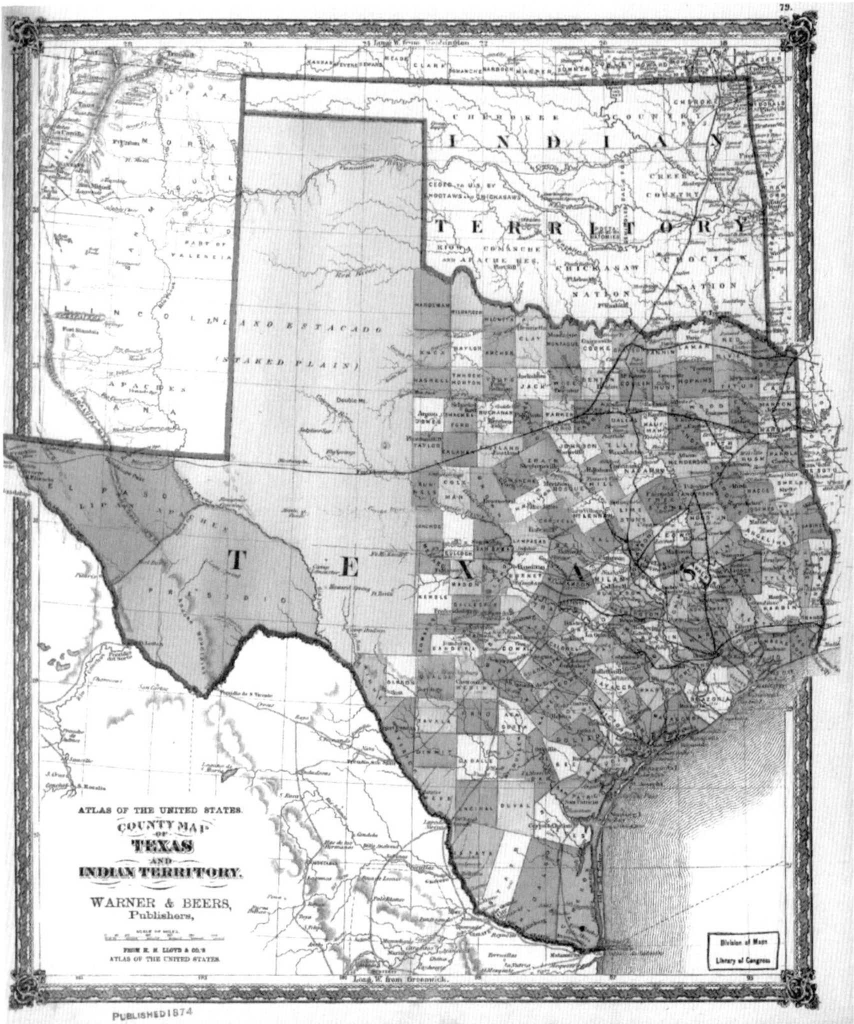

This Texas/Oklahoma map shows clearly the Butterfield Overland Mail route. *Public domain.*

Many station and Overland employees slept in large tents near the corrals, the corrals having been completed first. Indians were still a problem, although they had tempered a bit by 1858. Mules tended to be stolen at a regular rate, so wild ones were procured often.

Hitching wild mules to a wagon was a sight to behold. At one station, they overturned the mud wagon, ripping the canvas top. After losing a couple wild mules that refused to be harnessed, the wagon was eventually hitched, and the ride continued. Unfortunately, with some stations not completed, this caused a couple runs varying from thirty-five to seventy-five miles in length. Within a few months, however, all the stations were up and running.

Outside of Sherman, they crossed a "fine rolling prairie" covered with grass, but there were no trees and scarcely a shrub around. The first station after leaving Sherman was Diamond's, twenty miles away. They reached it in just three hours. Ormsby reported, "The beautiful moonlight lit up the vast prairie, making its sameness appear like the boundless sea and its hills like the rolling waves."

All along the way, they were greeted warmly as they entered each station. The meal and team preparations had to be done quickly since the stage was so ahead of schedule. At the station in Gainesville (founded in 1850), supper was thrown together hastily. The teams were changed quickly, and within a short amount of time, they were off again. The next station at which they arrived was fifteen miles away, and they made that distance in two hours and ten minutes. They were making excellent time.

The stations that were finished, such as the one at Connolly's on the edge of the Lower Cross Timbers, were built of coarse logs laid together rather roughly, the chinks filled in with mud. This one-room station was about twenty feet square—plenty room for two men. Here, as with most of the stations along the western route, breakfast consisted of a kind of shortcake, baked on the coals, and coffee. Sometimes bacon would be added.

Their ride continued across woods and plains to Jacksboro, which at that time was about one year old and contained a dozen or so houses. Upward of two hundred people lived around there, and due to the population, it was declared the county seat. They harnessed fresh mules here and rode all night through rolling prairie studded with mesquite.

Upon arriving at Fort Belknap, a frontier military station (established in 1851 and abandoned in 1867), on September 22 at 5:25 a.m., they were twenty-seven hours ahead of schedule. The fort was on the Brazos River and had a reduced number of cavalry assigned there. The town itself sported about 150 inhabitants and included several stores, a billiard saloon and a post office.

They crossed the low-running Brazos River, which at times peaked as high as sixteen feet, necessitating the use of a ferry. Forty miles farther, they crossed "sterile plainness" that was varied by clumps of black oaks and more mesquite.

At the Clear Fork of the Brazos, while the men changed teams, the passengers had a chance to dunk their tired, dust-encrusted bodies into the

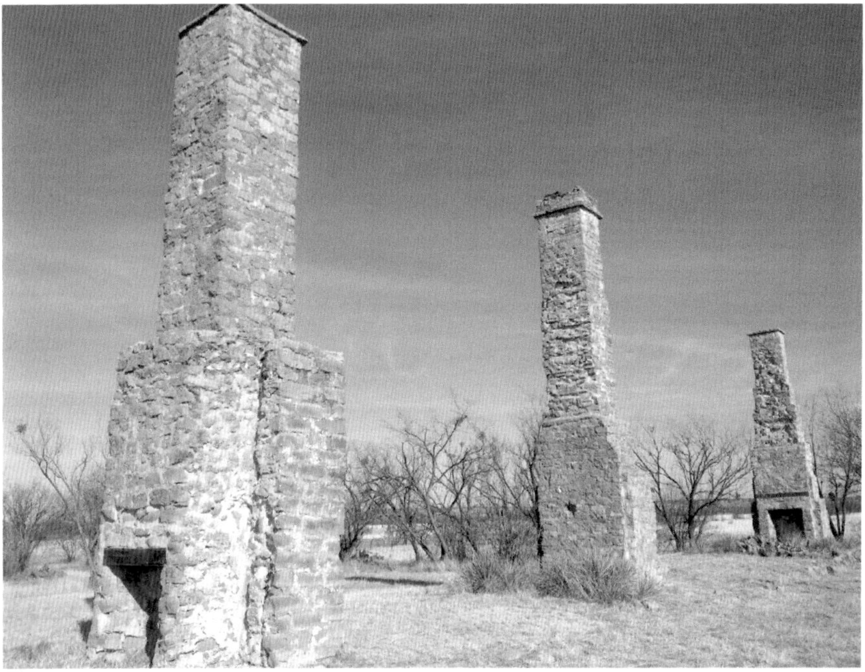

In 1853, the military abandoned Fort Phantom Hill. Shortly after the troops left, in April 1854, fire destroyed most of the buildings. *Public domain.*

water. But that sorely needed bath was over way too soon. Twenty-three miles later, they stopped at Smith's station, located on the banks of a small creek. No house had been built yet, but the corral, made of thick brush, was finished. Supper consisted of cake cooked in coals, clear coffee and dried beef.

Fort Phantom Hill was the next stop. The fort was established in 1851 and abandoned just three years later. The story goes that it was set afire by disgruntled soldiers as they rode away. All that was left standing were the chimneys, the ruins of the forty or fifty buildings now sitting eerily in the desert. The magazine, built entirely of rock, remained stable enough for the station manager to use as a storehouse. The stable was also in relatively good shape, as was another building used by the station manager and his wife. There was also a fine well, eighty feet deep and twenty feet in diameter, that had seventeen feet of water in it at the time.

The Overland Mail continued its journey with no major incidents. The mules remained cantankerous but managed to pull the wagons well. By the time they finally reached Fort Chadbourne (established in 1852 and abandoned in 1867) on Thursday afternoon, September 23—twenty-four hours ahead of schedule—they had traversed 955 miles.

The original route came north from Horsehead Crossing on the Pecos River, up the east side to Pope's Crossing and just below the New Mexico Border. However, eleven months later, the company switched it to the same route the San Antonio–San Diego line used because more water was available.

Not everything went as planned. About five miles from the Grape Creek station, at 2:00 a.m., Ormsby reported, "We came to a steep and stony hill… right in our path and impossible to avoid. One mule could neither be coaxed nor driven up, so we had to camp until morning when, after much difficulty, we ascended the hill and discovered the station…five miles distant."

The low rate of speed over this road was explained as "an entirely wild country across the great Staked Plain, seventy-five miles without water and one hundred and thirteen miles up the sierras, where stations have not yet been established. In fact, there is not a human habitation, except the company stations, on this whole distance of over four hundred miles."

They reached the head of the Concho River early on Saturday, September 25. The employees had not yet built a house and were living in tents, but they had made a large corral of bushes containing a large stock of mules. For breakfast, they ate broiled bacon, shortcake and coffee. Again, the meal didn't last long, and they were off to the Staked Plains. They arrived at Mustang Springs, where dams of water formed lakes or large pools, all spreading about three miles in length. Unfortunately, it was highly saline. But it was the only water for the next seventy-five miles, so they filled their canteens anyway.

Out on the Staked Plains, the road was hard and smooth, which made travel quick and easy. Mesquite bushes and yucca and other cactus lined the road. As they trotted across the plains, it "grew more dreary and the vegetation less thick." In fact, the only excitement they encountered in this entire seventy-five miles was a rattlesnake in the middle of the road. The driver stopped the team and took care of the snake, handing the rattles to the lone passenger, Waterman L. Ormsby.

Much later, the route turned into a hard-packed road full of stumps and bunches of weeds. Created by Captain Pope specifically for the Overland Mail Company, it left something to be desired. "The jolting was almost interminable and insufferable, and I frequently wished that Captain Pope could experience my ride over this road," stated Ormsby. "We pursued our weary course along the edge of the plain, thumping and bumping at a rate which threatened not to leave a whole bone in my body." He was not a happy camper at this point.

Monday, September 27, found them continuing their weary and dusty way up the Pecos River. The mules, ones they had to bring along since the stations were still too far apart, kept giving out, so they were left along the way. At least half a dozen ended up as wild mules in the desert. Finally, they arrived at Pope's Camp, where Pope had been drilling artesian wells. Unable to continue drilling, the project was eventually abandoned in 1859.

Here, the passenger and driver stopped long enough to get some supper—shortcake, coffee, dried beef and raw onions. A fresh team was harnessed, and again they were on their way. After running about thirty miles, they stopped near the head of the Delaware Creek, where they cooked breakfast over buffalo "chips." The meal consisted of jerked beef, raw onions, slightly wormy crackers and a bit of bacon. Apparently, they'd accidentally left coffee fixings behind at the last camp, so they had to make do without. Ormsby wrote, "The stomach, however, does not remain delicate after a few days of life on the plains."

The water at Independence Springs proved to be cool and refreshing, and pine trees grew in the gorge of the Guadalupe Mountains. At this station, in addition to the usual fare, venison pie and baked beans were added. Then, with a full stomach and fresh team, they headed for the Cornudas Mountains through Cañon de Guadalupe. Ormsby wrote, "The wild grandeur of the scene…is beyond description. The peak towers a full 600 feet above the base of the canon [sic], where is the tortuous, white sandy bed of a stream which was now dry but which in the rainy season must be a fierce torrent."

They got through about sunset, and the landscape made an indelible impression on the easterner. Just as the world turned dusky, Ormsby lay on the bed and closed his eyes. Voices he didn't recognize made him sit up. Robbers, perhaps? He leaned out and recognized the bright red stagecoach with "Overland Mail Company" emblazoned over the door. The eastbound Overland Mail stagecoach, which had departed San Francisco on September 15, had arrived! They were eight hours ahead of their schedule. Drivers and the five passengers exchanged greetings, congratulations, road information and bits of news, and then, like the proverbial ships passing in the night, each driver flicked the reins and the coaches went their own ways. September 28 would live in infamy.

The road from Guadalupe to the Cornudas was one of gradual ascent, with grass and mesquite bushes all around. The Cornudas are jagged rocks

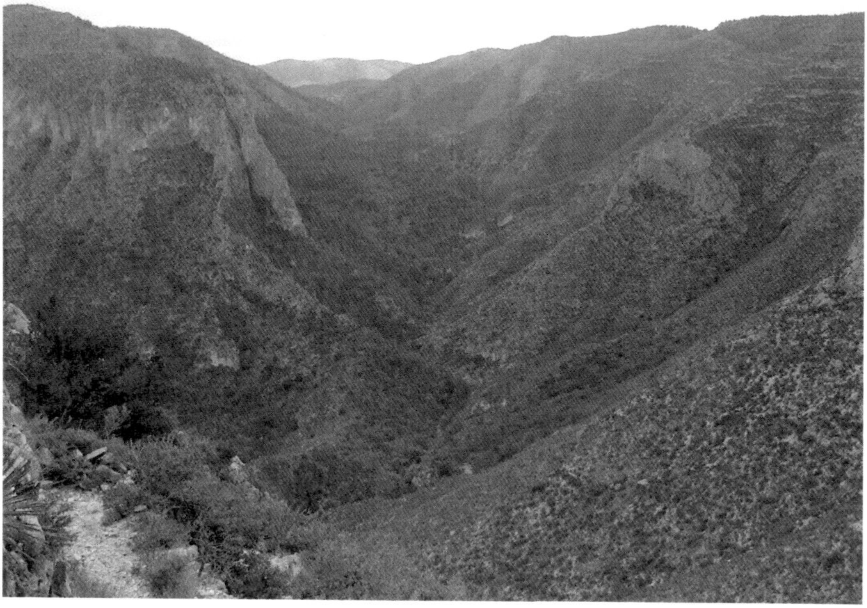

View of McKittrick Canyon from the Guadalupe Mountains. *Photo by Leaflet.*

jutting up about 1,500 feet from the plain. Water collects in natural basins that have been used by native Indians for hundreds of years.

At the Cornudas de los Alamos station, they changed teams, acquired a side driver and then set out for Hueco Tanks, thirty-six miles away. The road was in good shape, and the trip was easy. At Hueco Tanks, an excellent corral and station stood ready to assist the stage. Unfortunately, with the prolonged drought, water was not plentiful and had to be used sparingly for the animals.

Finally, the fertile valley of the Rio Grande came into view, the many vineyards and adobe ranches ushering the tired travelers into the tiny hamlet of Franklin (present-day El Paso). About two miles from Franklin was the adobe-built Fort Bliss, established in 1848. The station itself was located in a settlement called Magoffinsville.

Franklin, named after Benjamin Franklin Coons, who became the hamlet's first postmaster in 1852, was on the American side of the Rio Grande and

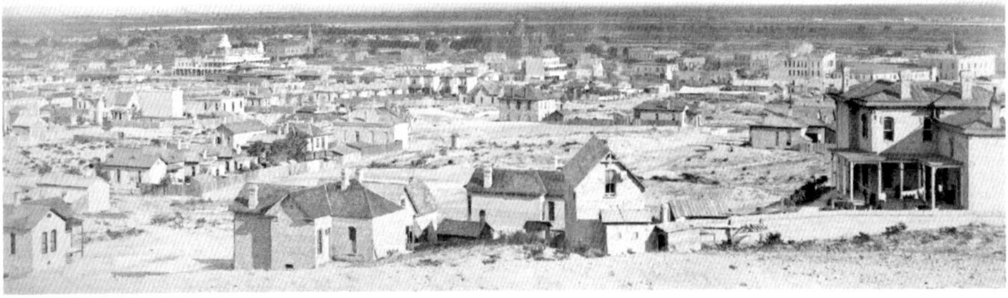

El Paso, 1880. In 2012, census records for El Paso recorded 672,538 people, making it the nineteenth-most populous U.S. city. *Photo by F. Parker.*

contained a few hundred citizens. What made the town—and especially El Paso del Norte (now Juarez, Mexico)—so appealing was its centuries-old use as a north–south crossroads. El Paso del Norte had served as a stopping point on the Camino Real, the road that came from Chihuahua and ended in Santa Fe. Later, with the discovery of gold on the West Coast, it changed axes to include east–west to the Pacific.

An earlier visitor traveling through in 1849 noted, "I am greatly disappointed in El Paso (Juarez)—it is a poor, miserable, dirty town, badly built." By 1850, Juarez claimed a population of about five thousand and was deemed a virtual paradise by most of the others traipsing across the desert. In July 1850, there were four thousand people with 1,200 to 1,500 wagons camped along the river both above and below the town.

Along with the influx of people—soldiers, teamsters, emigrants and merchants—came a host of problems. This far from any type of "real" civilization, people naturally took advantage of any form of law. The local farmers hoarded their food as more and more emigrants began to outnumber the actual residents. Any surplus food was sold for exorbitant sums. Many of the newcomers were "rough-hewn," and the consumption of homemade alcohol did not help to create law-abiding citizens.

So theft and bullying, plus a few gunfights—Mexicans against Americans and Americans against Americans—created a real sense of lawlessness, which became the trademark of the Wild West. Obviously, a market for anything that could be eaten, worn or ridden brought Mexican and American merchants here. Farmers left the East and moved to west Texas, and El Paso/Juarez blossomed. Correction, it exploded.

But not the American El Paso. The little cluster of adobe houses in 1857 contained no more than fifty people. There were twenty single-story buildings, most of which were vacant. In fact, in 1827, there was nothing on the Rio Grande's north bank, so Juan Maria Ponce de Leon, a well-to-do businessman from El Paso del Norte, received a grant for five hundred acres from the Chihuahua government for the total sum of eighty dollars. He built a fine home in what is now downtown El Paso in 1830. Anson Mills, a military man, made the first survey and plat for a town in 1859.

However, with the coming of transcontinental mail and the Overland Mail Company, the sleepy town changed. Franklin/El Paso quickly became the midpoint division headquarters and a merchant center. In 1857, a combination grocery store, saloon and billiard hall opened. Combined with that and the already established post office, Franklin became a convenient stop.

Today, most of the streets of the downtown section of El Paso—Overland Street, San Antonio Street, Santa Fe Street, Chihuahua Street and San Francisco Street—still reflect the direction of the old stage lines that came into or out of town. The American history of today's El Paso area really started with the Overland Mail.

Ormsby and the Overland Mail stagecoach left the town at 5:40 a.m. on Thursday, September 30, headed for La Mesilla, New Mexico. The trail left Magoffinsville, turned north and passed through today's Smelter Town, skirting Hart's Mill (created by Simeon Hart) and the future site of Fort Bliss. They followed the Rio Grande and entered New Mexico in present-day Anthony.

The road from Franklin north was quite hilly and dusty, although it followed the Rio Grande, which created a long green valley. Twenty-one miles north of Franklin, the Overland Mail changed teams at a pretty grove of cottonwoods. In fact, Henry Skillman, who drove that leg of the Overland trip, had a house there and used it as a home station.

Farther north sat Fort Fillmore, which was established in 1851 (abandoned in 1861) for the sole purpose of defending settlers against marauding Indians. Here there were quarters for officers and men, a good hospital, a magazine, storehouses, a bakery and a garden. A farm was also cultivated. Under the command of Lieutenant Freedley, eighty-three soldiers were stationed there. Prior to the Overland's arrival in 1858, there was already a combination mail station, gambling hall and whorehouse within walking distance of the fort. The Overland's station was located next to the fort's walls.

Later, in 1861, this fort, under the leadership of Lieutenant Isaac Lynde, would become the gateway for the Confederate army to charge

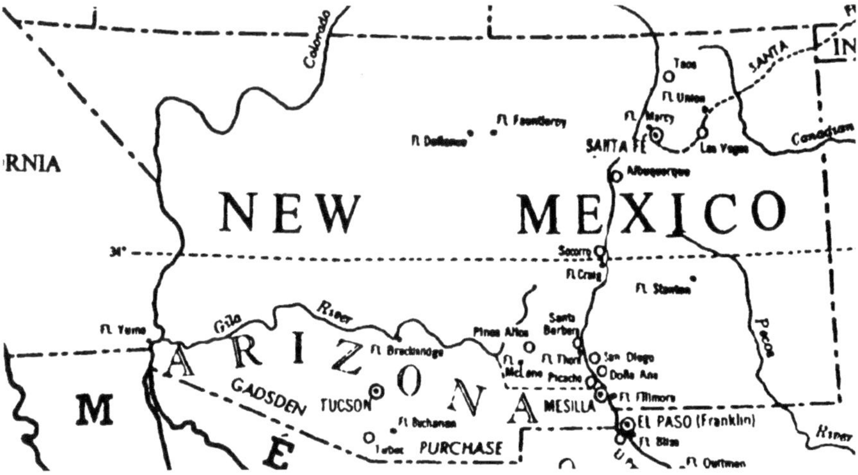

New Mexico and Arizona share much in common. From 1861 to 1865, the southern half was the Confederate state of Arizona, La Mesilla the capital. *Public domain.*

into southern New Mexico, plant its flag and declare the entire Gadsden Purchase area as the Confederate state of Arizona (1861–63). Today, Fort Fillmore is completely plowed under, and a pecan grove stands where soldiers once marched.

Approximately six miles after crossing the Rio Grande, the coach reached La Mesilla. The town was home to three thousand people, making it the largest settlement between San Francisco and St. Louis. While Orsmby had been excited to see this town, he was mighty disappointed when he got there: "Their houses were built on adobe and sticks…and in the city the houses are little better than on the outskirts…I never saw such a miserable set of people in my life." He went on to mention that several Yankee merchants were making a small fortune by selling lumber at $200 per one thousand feet.

La Mesilla, officially recognized by the United States in 1851, was established as a colony by Mexicans who did not want to become Americans when the boundary was changed following the Treaty of Guadalupe Hidalgo. However, in 1854, with the Gadsden Purchase, they became American citizens. A few moved across the Rio Grande, but most stayed put.

Street names in La Mesilla were nonexistent when Butterfield arrived. In order to clarify, the company chose to call the street where its offices were Main Street, which is now Calle de Principal. It then named the street where it dropped off passengers First Street, now Calle de Guadalupe.

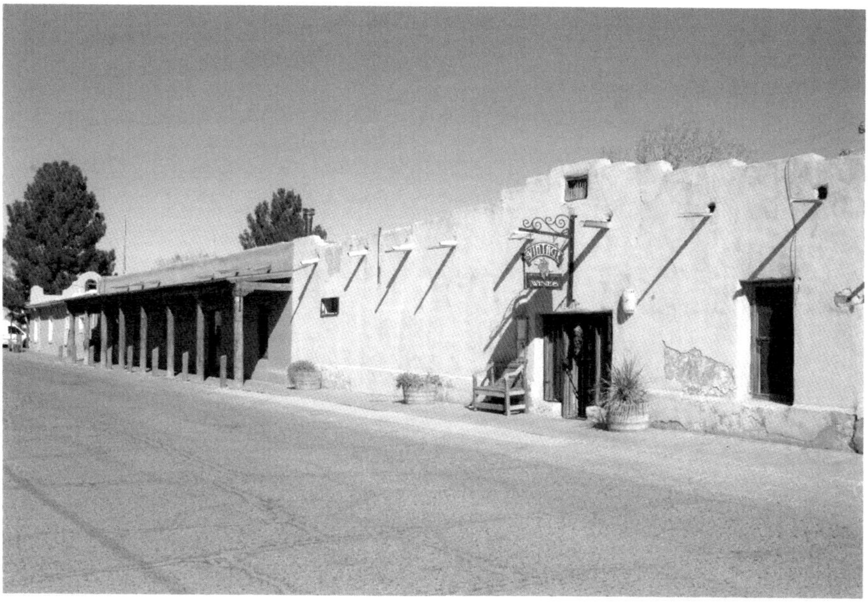

Wine and clothing shops now occupy what once was the Butterfield headquarters in La Mesilla, New Mexico. *Photo by author.*

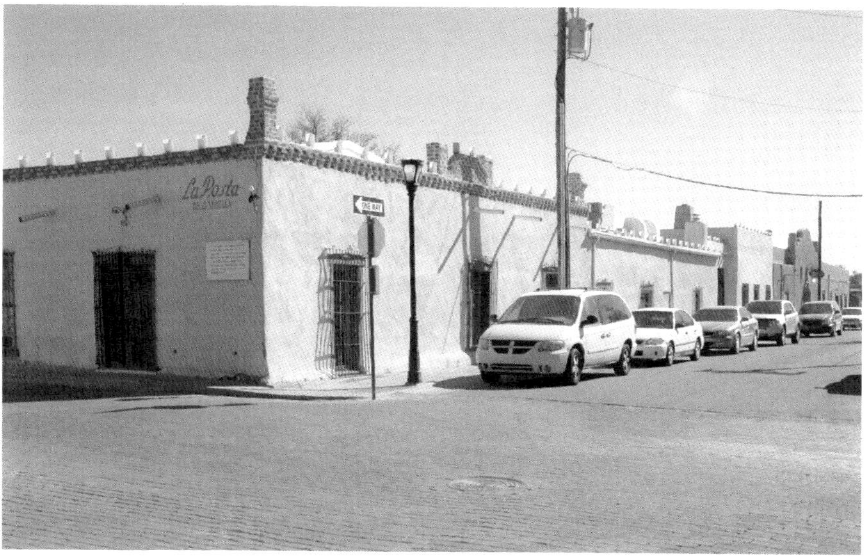

Considered the "backside" of La Mesilla's La Posta restaurant, this is where the Butterfield line dropped off its passengers. By 1870, it was home to the Corn Exchange Hotel. *Photo by author.*

Mesilla was another hub for stagecoach/freighting activity. On the north–south route between Chihuahua and Santa Fe, wagons rolled through town regularly, as did shorter stagecoach lines. This town was the destination and center of activities—social and otherwise. Emigrants from all over the world found themselves in this bustling town. Silver was mined in the Organ Mountains east of town, and since "the law" was few and far between, unscrupulous characters roamed as well. Dram shops (saloons) were opened on every corner and along the streets. In fact, today's El Patio, on the south side of the plaza, was opened as a dram shop by Sam and Roy Bean, who later became "Judge" Roy Bean of Texas. Somehow, it seemed fitting that the Overland divisional offices and drivers' quarters were on the same block as Sam and Roy's place.

Another dram shop a block over from the Overland office was the official "drop off" point for passengers. From here, they would either wait for one of three connecting stage lines or stay over to rest. This business grew from a saloon into the Corn Exchange Hotel in 1870, owned by Sam Bean, and today houses the La Posta restaurant. Inside, the restaurant gives homage to the building's colorful past by designating the original adobe room as the Corn Exchange Saloon.

A few miles north of Mesilla, the Overland changed from horses to mules and rounded Picacho Peak, where, due to a particularly steep patch, Ormsby walked almost two miles up a rugged hill through narrow Box Canyon. The reporter noted that when they reached the top, "we were upon the border of a broad and level plain extending far away as the eye could reach." Once back in the coach, they headed due west to Cooke's Spring, arriving there at 12:30 a.m. on October 1. Here they met the second mail coming from San Francisco. As an interesting aside, after this section of the route had been shut down, Cooke's Spring was the site of an 1861 ambush by Chiricahua Apaches in which Butterfield employees who were rounding up horses from the various stations were massacred.

By now, the stage had been traveling for seventeen days. They passed through New Mexico with no troubles, reaching Stein's Peak on the present-day Arizona/New Mexico border unharmed—tired, but in one piece. "The location of the station," Orsmby writes, "is in a little hollow under the mountain, so that we could not see it until we were within a hundred yards of it." They had heard a month or so earlier that a band of Apaches under Mangas Coloradas's leadership had gone to the station demanding twenty sacks of corn. The stationmaster saw fit to give it to the 250 waiting Apaches.

The Stein's Peak station was located in the eastern entrance to Doubtful Canyon, the colorful name ascribed to this pass because it was "doubtful" that travelers would make it through unscathed. The canyon is one mile from the Arizona/New Mexico border in the Peloncillo Mountains and named after Enoch Steen (mapmakers changed the spelling to Stein, but is pronounced "steen"). This station was started by Elias St. John (who was killed by Mexicans there) and finished in September 1858 by St. John's boss.

A reliable spring nearby provided year-round water. Like many other stations in that region, the Stein's Peak station was built of rock and bonded with adobe. Outside walls formed a seventy-foot- by forty-foot quadrangle. Most of that space was a corral. Walls were two feet thick and ten feet high. Large, flat stones formed the foundation, and the lower portion of the walls and stones got progressively smaller as it rose. This station was destroyed by marauding Apaches in 1861.

The Doubtful Canyon road west of the station was one mile of narrow wash, one and a half miles of gently rising broad meadow and three miles of deep rock-wall canyon. From there, it passed through an arroyo so narrow a wagon could not turn around in it.

Navigating through Apache Pass, considered the most dangerous portion of the entire 2,800 miles, was a real nail-biter. Following a meal of beef, bacon and shortcake, consumed in record time, Ormsby and the driver got back on the road and through the harrowing section as quickly as possible, especially since it was night and there was no moon. Ormsby wrote, "The pass…I could discern upon each side of the narrow road the perpendicular rocks, apparently ready to fall upon us. I was not sorry when we got through."

From Apache Pass, the road ran over relatively level ground, keeping more rugged mountains to the north. At Dragoon Springs, saguaro cactus, found only in southern Arizona, spike upward, standing sentinel over dry desert landscape. The route led through deep gullies and creek beds and stands of mesquite. Once across the San Pedro Valley, Ormsby wrote, "If there is any portion of the route calculated to impress one with the necessity of military protection…it is this very last few hundred miles which I have just described." However, he was quite taken with the beauty of the flora: "The land…is covered for the whole length of the river with…a large number of sun flowers…On the banks of the valley are some of the largest cactus plants I have yet seen." Of course, he's speaking of the saguaro.

The route took passengers and mail into Tucson. Ohio native Phocion Way wrote the following upon his arrival in Tucson in June 1858: "We were fully

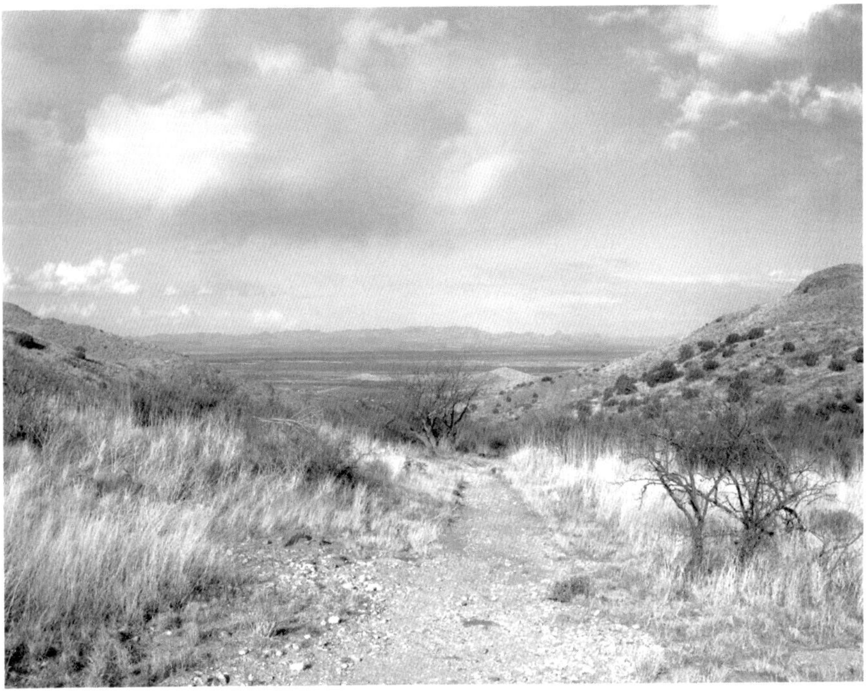

Apache Pass, Arizona, the site of two major skirmishes between Apaches and Americans, as it appears today. *Photo by Mark A. Wilson.*

prepared to see a miserable place and we were not in the least disappointed. There is no hotel or other accommodations for travelers, and I was obliged to roll myself in my blankets and sleep either in the street or the corral."

As dismal as that sounded, from the time Butterfield began operations in 1858, the town greatly improved. By March 1861, Tucson had become the center of transportation and supply while the population increased daily.

The initial Overland Mail arrived in Tucson at 9:30 p.m. on Saturday, October 2, and left half an hour later. At this point, they were 1,163 miles from the final destination of San Francisco. Orsmby stated that Tucson was a small town with mostly Mexican inhabitants. There were two or three stores kept by Americans.

From Tucson, the route then veered northwest to Picacho Peak, which would soon become the site of the westernmost skirmish of the Civil War. The road was laid out over a plain covered with sage, mesquite and cactus but no settlements of any consequence. Forty miles farther, at Maricopa Wells, situated on a large plain of alkali soil and coarse grass, they exchanged horses for a mule team. Forty miles farther, they arrived at the Gila River station at 9:30 p.m. on Sunday, October 3.

At the Dutchman station on the Gila, they took the opportunity not only for breakfast but also a sorely needed bath. After eating, they rode west along the river to Texas Hills, which was composed mainly of red slate rocks and, according to Ormsby, "look[ed] like a hill of icebergs" in the moonlight. By that evening, he had expected to get supper at the next station, but being ahead of schedule, the food wasn't ready. The driver insisted they push on, so everyone left hungry.

At a station twenty miles from Fort Yuma, they met the fourth stage out of San Francisco, thirty hours ahead of its schedule. It was waiting for Ormsby's stage in order to exchange coaches. The rate of speed on this division clocked at five and a half miles per hour.

Overland traffic used Jaeger's Ferry to cross the roaring Colorado River. Ormsby wrote, "We crossed the river at this point…The boat is a sort of flatboat, and is propelled by the rapid current, being kept in its course by pulleys running on a rope stretched across the river." On the other side, they were met with breakfast.

By now, the proverbial end was in sight. At least they were in the final state—only seven hundred more bone-jarring miles and they'd be in San Francisco. But for now, they ran along the banks of the Colorado, navigable for steamers of very light draught. The river continued to wear away the western banks, so much so that the road washed away with it. Quicksand patches were large and shifted, making travel difficult.

Ten miles from Fort Yuma was Pilot Knob, a mountain extending to the river, leaving but a narrow pass for the road. In order to escape sand dunes and a close-to-impossible road, the route turned south into Mexico, where three stage stations were manned—Cook's Well, Seven Wells and Alamo Mocho—before crossing back onto U.S. soil and into California. At one point, for several miles, the road lay directly in a gully. "The vegetation within sight was of but little account, while not a tree could be seen except upon the distant mountains," noted Ormsby.

The route then crossed into the Imperial Valley and turned northwest through foothills and up a broad valley to Vallecito and then Warner's Ranch, which, according to Ormsby, was a comfortable house situated in the valley in the midst of a beautiful meadow, with its shingled roof looking "more like civilization than anything I had seen for days."

Now well out of desert country, the route consisted of smooth valleys among rocky mountain ranges. Today, a modern highway runs directly on

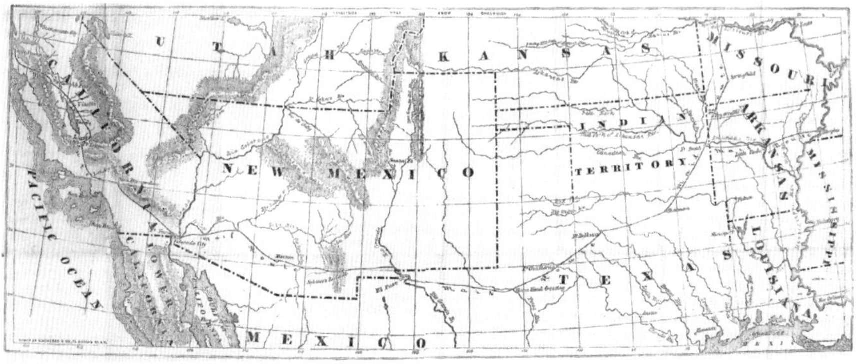

The name "California" is said to originate from Garci Montalvo's 1510 story about a fictional paradise peopled by black Amazons and ruled by Queen Calafia. *Public domain.*

top of the old trail extending through this mountain scenery. Farther up the road, they passed oak groves and springs and stopped at the stations of Hall's Oak Grove, Swango, Laguna and Temecula. At this point, other stage lines connected with San Diego and San Bernardino. On April 22, 1859, because of Butterfield's mail line, the first inland Southern California post office was established in Temecula in the Magee Store where Butterfield's Overland stage stopped. This was the second post office in the state, the first being located in San Francisco.

The route continued along the Santa Ana Mountains south of San Bernardino and close to Los Angeles. This ran it directly through El Monte, the oldest strictly American settlement in Los Angeles County and what many considered the end of the trail for emigrant wagon trains. Clearly, Ormsby was glad to be in California: "The post office is a wooden building, with a neat piazza and shingled roof, which looked quite refreshing after over a thousand miles' travel without seeing a house having the appearance of civilization."

Los Angeles, incorporated in 1850, was about twenty-five miles up the Los Angeles River from the bay. Its full name was a mouthful: Pueblo de Nuestra Señora la Reina de los Angeles de Porciúncula. Census records in the 1860s report a total population of 4,385, mostly Mexican, Spanish and Indian. A fertile strip of land connected the city and lowlands of the ocean and had been a Spanish and Mexican settlement since 1777. In fact, the first California gold was discovered in the San Fernando hills about thirty miles from there. When the Overland began running through there, the main reputation of Los Angeles was a cattle center for the state.

"We arrived at Los Angeles in five days and four hours from Tucson, making nearly six miles an hour on the average, in spite of the sandy desert and craggy hills," Ormsby writes. However, they stopped just long enough to change coaches and hit the road again. The fact that the Overland Mail Company was the first direct connection with the East set the stage for Los Angeles to become a major transportation center.

Continuing north through the San Fernando Pass, the route passed through Tejon Pass of the Sierra Nevadas. Ormsby says of this pass, "I believe, the steepest hill on the whole route…it's certainly a very steep hill and our six horses found great difficulty in drawing our empty wagon up."

Despite bad road conditions due to a recent severe rainstorm, they made the trip to Hart's ranch, eight miles away, at nearly ten miles per hour. From here, they rode through twelve-mile-long Francisquito Cañon and into Fort Tejon (established in 1854 and abandoned in 1864), ninety-six miles from Los Angeles, at 3:40 a.m. on Friday, October 8.

At Fort Tejon, hay was forty-three dollars per ton and barley six cents a pound, a surprisingly exorbitant cost, Ormsby noted. Back home, the same would run a tenth of that. The stage left at 4:33 a.m., time allowed for breakfast and a stretch.

About one hundred miles farther north, the route took passengers and mail through the growing town of Visalia, the first town after Fort Tejon and the first settlement of any importance since Los Angeles. Here, most of the countryside was settled and the road relatively defined. "The place is situated in an oak grove, sometimes called the Four Creeks, which seldom have much water. There are a number of New Yorkers in the town, and I took a glass of lager, which made me feel quite at home," Orsmby reported.

News of the stage's arrival spread quickly, although it was nearly midnight when they arrived. "Before we left, they gave us an anvil salute, which was quite new to me," wrote Ormsby. In an anvil salute, still done today, gunpowder is placed in a hole in one anvil, a fuse is laid across the powder and another anvil carefully placed over it. When the fuse is lit, the one-hundred-pound steel anvil shoots upward, arcs and then crashes to earth. "This was the first evidence of enthusiasm along the route since we left Fort Smith."

From Visalia, the route crossed the San Joaquin River to the stage stop of Cross Creek, thirteen miles from town. Farther up the road, they stopped at King's River, Elk Horn, Fresno City and then Firebaugh's Ferry. From there, the trail continued alongside the river for forty miles until it reached San Luis ranch, the only house within thirty miles, at the entrance to Pacheco Pass. We can picture Ormsby almost smiling when he wrote, "The owner's hospitable

table is always open to all who pass that way. It is a great rendezvous for drovers going down into the valleys after cattle."

The distance through Pacheco Pass was twelve miles of harrowing, breath-sucking awe. "On every side we could look off down steep and craggy ravines, some of whose bottoms could not be discerned in the distance. Our road led immediately on the brink of many a precipice, over which a balky horse or a broken axle or an inexperienced driver might send us whirling in the air in a moment." Ormsby was invited by the driver, Tote Kinyon (the brother of one of the Overland's owners), to sit up on the bench—quite an honor in stagecoach days. What he got was the ride of a lifetime. The twelve miles through the canyon required outstanding driving skills. Kinyon, in a hurry with the "first States' mail," put the pedal to the metal, so to speak. "I expected to see him put down the brakes with all his might, but he merely rested his foot on them, saying, 'It's best to keep the wheels rolling, or they'll slide'; so he did keep the wheels rolling, and the whole coach slid down the steepest hills at the rate of fifteen—yes, twenty—miles an hour, now turning an abrupt curve with a whip and crack and 'round the corner, Sally,' scattering the loose stones, just grazing the rocks, sending its rattling echoes far away." He thought, "If I…was destined to be killed in a stagecoach I most certainly should have considered my time come."

The wheels remained on the coach, the horses on the road and Ormsby and the driver in their seats.

Once safely through the pass, they continued for two hours on to Gilroy, where they ate supper and changed teams. While they weren't met with an anvil salute, they were bombarded with questions. "What's the news from the States? How did you like your trip? Have plenty to eat?" And on and on. Gilroy was a flourishing town boasting over six hundred citizens, a number of very fair houses and several stores.

Thirty miles from Gilroy, the coach made its way through the swampy reaches of San Francisco Bay and then headed northwest to San Jose, a well-established community dating to 1777. Originally named El Pueblo de Jose de Guadalupe, like Los Angeles, it grew into a shortened name. By 1849, it was the temporary territorial state capital until 1851, when that title was given to Sacramento. It was here that passengers and mail shifted from the durable Celerity mud wagons into the dashing Concords for the final leg into San Francisco.

Although it was an old settlement, established long before the Mexican war, San Francisco in the 1850s was a bedlam of shacks, tents, a few buildings, masses of rotting wooden ships and signs in a variety of languages—English,

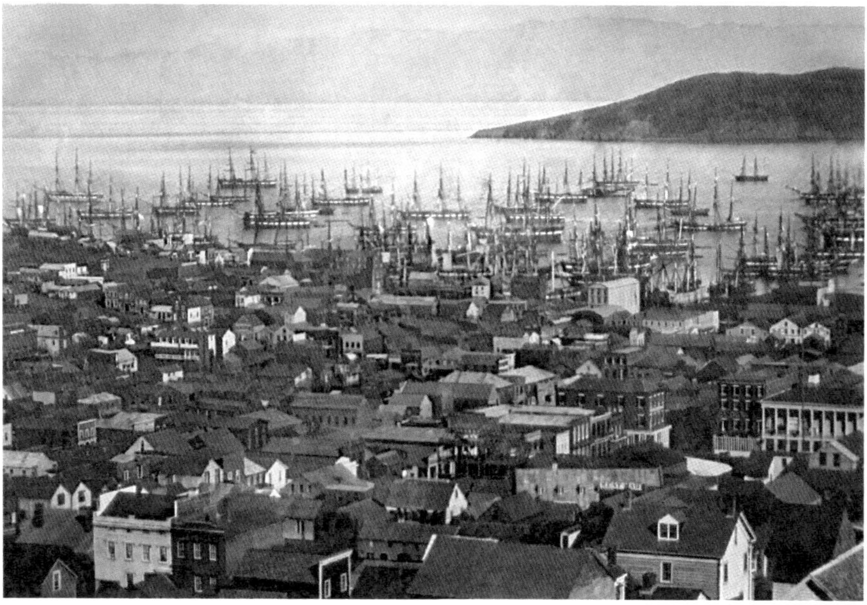

San Francisco Harbor, pictured here in 1851, would become so crowded that ships often had to wait days before unloading their passengers and goods. *Public domain.*

Spanish, Chinese, German and French. Violence and death were everyday occurrences as new arrivals would find themselves bossed by tough bands of thieves and killers imported from New York's Bowery district. The Barbary Coast of San Francisco history was fierce and legendary. To add to the mix, newspaper editorials urged formation of vigilante committees, taking the law into their own hands.

Companies of armed citizens, strong enough and armed enough to defend themselves, patrolled the streets, staged raids and filled the rafters with hanged culprits. By 1855, most of the criminal element had been eliminated. By the time of the Overland Mail in 1858, the roughest toughs were gone, but San Francisco retained frontier settlement in the form of bear fights, dogfights and gambling dens. On the other hand, there were scores of churches, schools and mercantiles. And more permanent wooden structures took the place of tents. A civilized town emerged along with the connecting of the United States.

Ormsby reported, "It was just after sunrise that the city of San Francisco hove in sight over the hills, and never did the night traveler approach a distant light, or the lonely mariner descry a sail, with more joy than did I the city of

San Francisco on the morning of Sunday, October 10. Swiftly we whirled up one street and down another, and round the corners, until finally we drew up at the stage office in front of the Plaza, our driver giving a shrill blast of his horn and a flourish of triumph for the arrival of the first overland mail in San Francisco from St. Louis…The mails must be delivered, and in a jiffy we were at the post office door, blowing the horn, howling and shouting for somebody to come and take the overland mail."

But nobody came. Not until a half hour later at 7:30 a.m., when a yawning, bleary eyed postal worker opened the doors. He was probably startled to see the Overland Mail Company stagecoach waiting outside. Finally, the mail was delivered. It was a rather lackluster ending to an awe-inspiring endeavor that changed the world. "And I had the satisfaction," Orsmby writes, "of knowing that the correspondent of the *New York Herald* had kept his promise and gone through with the first mail—the sole passenger and the only one who had ever made the trip across the plains in less than fifty days." It actually took him only twenty-three days and twenty-three hours.

Quite an accomplishment, indeed.

Ormsby spent a few days in San Francisco touting stage travel. Even though it had been fatiguing, he said, he'd be willing to repeat it. However, he took a steamer home.

Chapter 23

A QUICK LOOK AT JUST HOW FAST THEY TRAVELED

Rate of Speed from Point to Point (East to West)

Trip	Miles Per Hour	Mileage	Hours
Tipton to Fort Smith	5 ½	318.5	48.55
Fort Smith to Sherman	5 ½	192	38
Sherman to Fort Chadbourne	3 ¾	282.5	65.25
Fort Chadbourne to Franklin	2 2/3	458	126.3
Franklin to Tucson	9	360	82
Tucson to Los Angeles	5	562	143.65
Los Angeles to San Francisco	9	462	80

TOTAL: 2,795 miles, 575 hours (23 days, 23 hours)

Chapter 24

THE TRIP HAS ENDED, BUT THE COMMENTS ARE JUST BEGINNING

Comments handed down through the decades by unnamed passengers tell a story that only they could tell:

One rider proclaimed, "Were I to make the trip again, I would take the stoutest suit of clothes I could get, with a strong, loose pair of boots."

Another noted, "The fatigue of uninterrupted traveling by day and night in a crowded coach, and in the most uncomfortable positions, was beginning to tell seriously upon all the passengers, and was producing in me a condition bordering on insanity."

A Tucson judge said of the Overland stage: "It minded not God or man. Its only aim was to get through on time."

The January 4, 1859 *San Francisco Bulletin* reported, "The towns of Mesilla and El Paso have few attractions. The adobe houses look dirty and the people are negligent and idle. Cottonwood trees stand near most of the buildings and serve to somewhat relieve the appearance of these towns."

G. Bailey, a passenger on the first eastbound stage out of San Francisco, reported, "The various difficulties of the route, the scant supply of water, the long deserts, the inconvenience of keeping up stations hundreds of miles from the points from which their supplies are furnished; all these and the minor obstacles, naturally presented to the successful management...have been met and overcome by the energy, the enterprise and the determination of the contractors."

TIMES AND DISTANCES FROM BAILEY'S REPORT (WEST TO EAST)

ROUTE	MILES	HOURS
San Francisco to Los Angeles	462	80
Los Angeles to Fort Yuma	282	72.20
Fort Yuma to Tucson	280	71.45
Tucson to Franklin (El Paso)	360	82
Franklin to Fort Chadbourne	458	126.30
Fort Chadbourne to Colbert's Ferry	282.5	65.25
Colbert's Ferry to Fort Smith	192	38
Fort Smith to Tipton	318	48.55
Tipton to Saint Louis (via rail)	160	11.40

TOTAL: 2,795 miles, 596.35 hours (24 days, 18 hours)

Chapter 25
TO SUM IT ALL UP

To many people who think more of their ease than they do of physical health, a stage ride of a thousand or two thousand miles may seem a very formidable undertaking. But for those who have a liking for adventure, and a desire to see something of the world, a long ride of two or three weeks, practically in the open air, possesses a wonderful charm, especially in remembrance.

—H.D. Barrows, 1860s stagecoach passenger

PART V
THE END OF AN ERA

John Butterfield's vision of connecting the United States, providing mail delivery at a reasonable time and making people's lives better came to an end sooner than anticipated. Yes, everyone knew the telegraph would eventually reach coast to coast. And the train? That had been a foregone conclusion as early as 1850. So while it was no surprise that the great Overland Mail Company lost its luster, it was surprising how quickly it happened.

An unbelievable series of events led to the ending of an unprecedented linkage of the United States. Butterfield's recall as president of the company, his subsequential ill health, the Bascom Affair, the expansion of the Pony Express, Indian uprisings, the telegraph and the Civil War all played major parts in the Overland Mail Company's demise.

Chapter 26
BUTTERFIELD IMPEACHED/ STEPPED DOWN/REASSIGNED

When John Butterfield formed the Overland Mail Company, he was not a young man. In those days, a man of fifty-six was considered "old." In fact, Butterfield was known as "Old John." In 1860, at the age of fifty-nine, Butterfield had a small stroke that caused him to cut back on his duties. It was also known that the Overland had lost money, the company costing more than $1 million a year to maintain. However, it had never been touted as a moneymaking endeavor. Butterfield's intent was to strengthen the nation through connecting it with information. Communication was the key, he felt.

A couple of the company's board members, in particular Henry Wells and William Fargo (of Wells Fargo fame), disagreed with most of what Butterfield said and did. They and William Dinsmore staged somewhat of a "coup" and nudged Butterfield out.

"Old John" politely and officially "stepped down" as president of the company but remained on the board. Immediately, morale and efficiency suffered. William Dinsmore took the helm, but he wasn't the man John Butterfield was.

It was reported on January 4, 1861, in Tucson: "The changes made in the management of the Overland Mail route have materially retarded the speed of the coaches and worse time is now made from Tucson in both directions than at any other former period. It will require but a short time to demonstrate, to the entire satisfaction of the directors, that the policy which dictated this change was unwise." A report out of Washington stated,

"...the Butterfield Overland Mail Company is now under the control of Wells, Fargo & Co." There was speculation Butterfield wanted to start a Pony Express operation on the southern route, and the board opposed it. Wells Fargo took over the remains of the company and set up a mail-passenger service between Salt Lake City and Placerville, California, using the salvaged equipment and contract.

It could have also been that Butterfield opposed his partners' desire to haul money, gold or express (valuable equipment) in order to discourage robbery. In January 1861, the *Mesilla Times* reported, "Well & Fargo [*sic*] have opened an express between Arizona and California. The termini of the two divisions of the overland mail are to be located at Mesilla—the eastern going to Fort Davis, the western to Tucson."

We do know that William Dinsmore was a partner in the Wells Fargo Express Company, and it's possible he pushed to oust Butterfield. At any rate, quite a bit of political maneuvering took place.

Maybe Butterfield acquiesced to a plan to move the route north, as his partners dominated that route at the time he struggled to keep his southern route going. What is clear is that his partners profited by working against him and in competition with his southern route.

It may never be clear why this takeover happened, but it certainly changed the face of the Overland Mail Company. Once Butterfield "stepped down," morale sagged.

John Butterfield returned to his native New York and lived a productive and active life. While on a business trip to New York City, he suffered a major stroke in 1867 and returned to Utica in his private railroad car. He died in Utica on November 4, 1869, nine years after being pushed out of the Overland Mail Company. His widow died in 1883.

Chapter 27
THE BASCOM AFFAIR

It usually takes only one bad apple to spoil the entire crop. In this particular case, there were two bad apples—one a nefarious rustler who started the whole thing, and the other a hardheaded military upstart. Up until February 1861, Apaches and Americans got along reasonably well in Arizona and the Apache Pass area. John Butterfield had worked hard to keep peace with the people who lived there. He had traded goods with them, bought firewood from Cochise and kept his distance when asked. In fact, there were no hostilities or war paint applied until a ne'er-do-well drunkard named John Ward set up business and picked up a contract to sell beef to Butterfield's station at Apache Pass and nearby Fort Buchanan.

Fresh water in the desert is always a controlling factor, and it was no different for the Chiricahuas and travelers of the 1800s. So when Butterfield met with Cochise over "water issues" in 1857, the agreement was reached that everyone could use it peacefully. In fact, Cochise saw the Americans as becoming allies against his hated Mexicans. And the fact that the Apache Springs and the new stage station were situated right in the heart of Chiricahua territory created a potential burbling pot of menace for both sides. However, they got along—until John Ward entered the picture. A "cattleman" of questionable character, Ward supplied meat from any beef that happened to wander by, either coerced or accidentally. Ward, an Irishman who had migrated to Arizona, lived with a Mexican mistress, Jesusa Martinez, and her ten-year-old half-breed son, Felix. Together, Ward and Jesusa had five children before his death in 1867. Felix's real father was

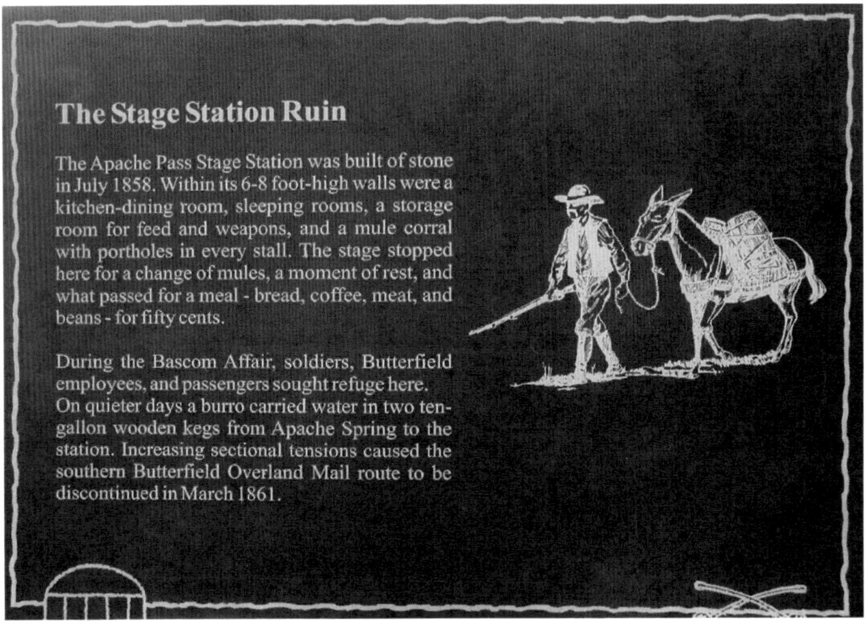

Apache Stage Ruins plaque. *Photo by Myke Groves.*

said to be an Apache warrior (who was also part Irish with the last name of Telles) who had captured Jesusa. Three years later, he went off to raid somewhere and never returned.

Ward went on rampages of his own, thanks to excessive bouts with alcohol, and when the mood came, he would unmercifully beat the boy. In October 1860, after one particularly brutal beating, Felix ran away at the same time another band of Apaches, the Pima, ran off John Ward's herd. When he sobered, Ward became an indignant but oh-so-righteous citizen and complained to the commanding officer at Fort Buchanan. How dare the Indians steal his cattle! Oh, and by the way, they had apparently stolen the kid, too. He blamed the Chiricahuas.

Citing short-handedness, the staff at Fort Buchanan did not pursue the issue until a young lieutenant, fresh out of West Point (graduating twenty-sixth out of twenty-seven), was sent west, where he chose to prove himself by returning young Felix Ward to his rightful family. George N. Bascom was eager to make a name for himself, and history shows that he did just that—but it probably isn't the name he was seeking.

On February 4, 1861, Second Lieutenant George Bascom and his detachment of fifty-four men encamped at Siphon Canyon, a few miles away from and within view of the stage station. After establishing camp, Bascom sent out word that he'd like to talk with Cochise. Contact at that time was simple, as the springs served as a communication center for the entire area. Soon Cochise; his brother Naretena; his wife, Nahlekadeya; his young son, Nachise; and two warrior nephews gathered for the social call. Little did they know that Bascom had other ideas.

Bascom's mission was to find Cochise, recover a kidnapped boy and return livestock assumed taken by the Chiricahua Apaches. By now, it was known that the Pimas, not the Chiricahuas, had stolen the cattle, but Bascom failed to take that fact into account.

At the camp, the Indians were invited inside the extra-large tent that Bascom had set up as a snare. They all sat down for polite chitchat. Outside, soldiers surrounded the tent, ready to spring at the prearranged signal.

During the meeting, Bascom accused Cochise of stealing the cattle and boy. Cochise, known for his strict adherence to the truth, was astonished at the accusation. "What boy? What boy?" he asked through an interpreter. When he finally understood what Bascom demanded, Cochise promised to look into it. He explained to Bascom that neither he nor his tribe knew anything about it and that he had come to the meeting of free will and brought family as a measure of good faith.

Undaunted and ignoring the facts, Bascom accused Cochise of lying. Cochise, understandably, grew extremely irate. Bascom stated indignantly that Cochise and his people were prisoners until the boy and livestock were returned. Even if they had taken the cattle, it was a safe bet that all evidence was gone, now in the form of beef jerky and stew. This accusation infuriated Cochise, who rightfully expected Bascom to know that Apaches took cattle only for food. The additional challenge to his honesty sent Cochise over the edge. He drew a knife from his legging, sprang to his feet and ripped a slit in the tent. He then plunged through the opening, bowled over a soldier and disappeared into the canyon.

Immediately, the soldiers drew guns, and the remaining Apaches were captured. Cochise retreated into the hills and collected a small band of his men. Upon reflection, Cochise decided that his friend James Wallace, a stage driver, was part of the deceit. Cochise called out to the men at the station. Wallace, Charles Culver (the station agent) and Walsh (a hostler), unaware of what had just happened, stepped outside to talk.

Cochise and his men attacked. In the ensuing fight, Wallace was captured, Culver shot by the Apaches and Walsh shot by a couple soldiers who were aiming at the Indians. With only Wallace alive, Cochise had what he thought would be bargaining power to have his family returned.

Bascom, refusing to listen to reason and the voice of his subordinates, who knew for a fact that Cochise was innocent, hanged the nephews. However, he did release Cochise's son and wife.

According to Sergeant Daniel Robinson, who was stationed at Fort Buchanan, "Our wagons were placed end to end forming a semicircle, covering one side of the station, and the corral, making a basis for outer breastworks. There was a deep ravine on this side, the head or nearest point of it about a hundred yards from the station. Empty grain sacks were filled with earth and placed on the inner side of the circle."

Justifiably outraged, Cochise ordered full-out war. Now the Americans were just as much enemies as the Mexicans. In retaliation, Cochise and his band lassoed Wallace to the back of a horse and dragged him to death. On solider later recalled, "…in a small valley on the western side of the Chiricahua range we came upon three bodies, one of which, upon examination, I knew to be that of Wallace by the gold filling in some of his teeth. All of the bodies were littered with lance holes."

The relationship between the Apaches and Butterfield would never be the same. In the days following the mishap, Cochise ordered all wagons, travelers and stage stations in his jurisdiction to be killed or burned. Since his area ran from west of Tucson to the Rio Grande, it was hundreds of square miles of mayhem. Bascom's rash actions had ignited a major Indian war.

The Chiricahua killed and mutilated at least 150 Americans within the next two months alone, and the friendship turned into a campaign to drive out the Americans.

Only one Overland stage was ever attacked, and that happened on February 16, 1861. It was well known that the Butterfield stage kept to a strict schedule, so Cochise and his men laid a trap in the pass. They placed tumbleweeds and brush in the road, causing the stage to stop. They had planned to swoop down and attack; however, as the westbound stage came through early, the driver and passengers simply got out, cleared the way and went on, unaware of the planned ambush.

It wasn't until hours later when the eastbound passed through, shots rang out and two horses went down. The passengers kept up a covering fire while A.B. Culver (station agent Charles Culver's brother) and another man cut the dead animals from the harness. They continued and fought their way to the safety of the station.

The Indians satisfied themselves with raiding the horse corral and then left. The station was never attacked. As far as the Overland was concerned, this was the final action in a very messy affair.

In all of this, the Overland was not at fault. It was simply a company caught between military and Apache forces. Unfortunately, two Butterfield employees lost their lives at the expense.

For the next twenty-five years, travelers throughout the Southwest did not enjoy the peaceful passage experienced during the Butterfield days.

Chapter 28

AND NOW . . . THE REST OF THE STORY

George Bascom, the wrong man in the wrong place at the wrong time, had caused or sped up the rise of the Apaches. He died on February 21, 1862, almost a year to the day of the mishandled Apache affair, during the Civil War fight at Valverde on the Rio Grande in New Mexico. He commanded the rank of captain by then. As an aside, Dr. Bernard J.D. Irwin, a military medical officer who arrived in the middle of the affair, would later go on to receive the first Congressional Medal of Honor for "going to the aid of Lt. Bascom at Apache Pass." Irwin led fourteen men of the First Dragoons to rescue Bascom, who had been besieged by Cochise. Irwin then tricked Cochise into thinking there were many more soldiers around, and Cochise withdrew his men. Irwin and Bascom joined forces and together found the boy in the mountains. (Dr. Irwin was promoted to brigadier general in 1904.)

And the young boy who ran away? He grew up to be known as Mickey Free and worked as a scout for the U.S. Army, serving George Crook. Later, he became a bounty hunter. He spoke English, Spanish and a couple Apache dialects. His gray eyes, one of which had been blinded in a hunting accident when he was a teenager, were unusual in that part of the country. Free had joined the Apache scouts in 1872 and within two years had been promoted to the rank of sergeant. He was eventually promoted to first sergeant and appointed to Camp Verde to serve as an interpreter. As a bounty hunter, Free tracked the Apache Kid, who had a $15,000 reward on his head. Mickey Free died in either 1913 or 1915.

Cochise was described as a large man with a muscular frame, classical features and long black hair, which he wore in traditional Apache style. He was about six feet tall and weighed about 175 pounds. In his own language, his name "Cheis" meant "having the quality or strength of oak." He prized honesty and truth above all else.

Following various skirmishes and his uprising beginning in 1861, Cochise and his men were driven into the Dragoon Mountains. Nevertheless, they used the mountains for cover and as a base from which to continue attacks against the white settlements. Cochise managed to evade capture and continued his raids against white settlements and travelers until 1872, when a treaty was finally negotiated by General Oliver O. Howard with the help of Tom Jeffords, Cochise's only white friend.

After making peace, Cochise retired to his new reservation (with his friend Jeffords as agent), where he died of natural causes (possibly abdominal cancer) in 1874. He was buried in the rocks above one of his favorite camps in the Dragoon Mountains, now called Cochise Stronghold. Only his people and Tom Jeffords knew the exact location of his resting place, and they took the secret to their graves.

Nachise (alternately spelled Naiche, Natchez or Nache), which means "meddlesome one" or "mischief maker," the youngest son of Cochise and his wife, Dos-teh-she, lived from 1856 to 1919. He was a tall, handsome man with a dignified bearing that reflected the Apache equivalent of a royal bloodline. His mother was the daughter of the great Warm Spring/Mimbreño chief Mangas Coloradas.

In the 1880s, Nachise and Geronimo successfully went to war together. In Oklahoma and, later, New Mexico, Nachise was reputed to be the finest Indian artist of that period. He painted his color pictures on deerskin. His subjects were flowers; deer, turkey and other wild animals; and various objects of nature. He also carved canes from wood and painted them in different colors. Nachise eventually moved to the Mescalero Indian Reservation in New Mexico, where he died in 1919.

The Bascom Affair changed everything. Following the affair, the War Department pulled all troops out of Arizona and moved property, supplies and livestock back into New Mexico in preparation for a transfer of everything to the northern states as the threat of civil war grew. With fewer soldiers around, the Indians increased their attacks on civilians and civilian operations (including other mail and express services), as well as innocent emigrants.

Chapter 29
THE PONY EXPRESS

Iconic images of slightly built male orphans galloping across the plains, their mail bags flapping against their steed's side, have been etched permanently into what people think of as the classic American West. Yes, they did blaze across the central United States with speeds averaging twenty miles per hour. Yes, they employed older boys and men of small build. And, certainly yes, "the Pony," as it was known, made a noteworthy mark in history.

As Mark Twain wrote in *Roughing It*, "So sudden was it all [the sight of the rider], and so like a flash of unreal fancy, that but for the flake of white foam left quivering and perishing on a mailsack after the vision had flashed by and disappeared, we might have doubted whether we had seen an actual horse and man at all, maybe."

But the story behind the Pony Express intertwines with Butterfield's Overland Mail Company like string around a yo-yo. Running for eighteen months and carrying mail from St. Louis, Missouri, to Placerville, California, the Pony Express lost money. A lot of money. But so did Butterfield's Overland Mail Company. "The Pony" carried only letters, no passengers, in stagecoaches. It was never designed to do anything but transport important letters and documents across the middle part of the continent. And it did this quite well.

The one-way trip was supposed to be made in ten days in the summer and twelve in the winter, the route going over the Sierra Nevadas. First undertaken privately and then by the government, it ran from April 3, 1860, to October 24, 1861. But the idea of a pony express wasn't new—versions of

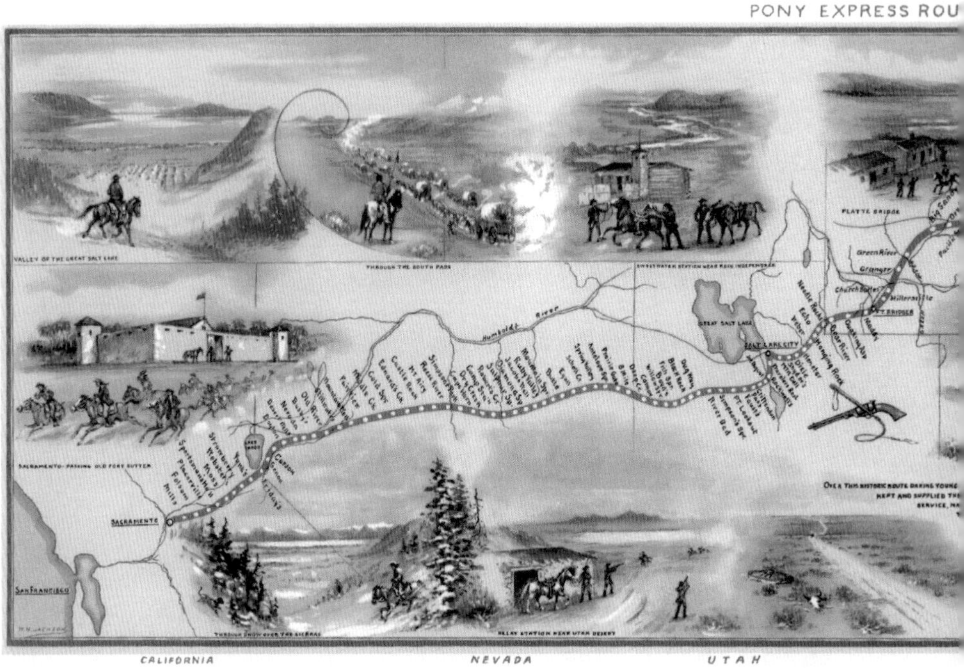

The first Pony Express rider was either Johnson William Richardson or Johnny Fry. "The Pony" ran for eighteen months. *Map by William Henry Jackson.*

the long-haul express operated as early as 1852. These feeder pony express routes were run by various individuals and express firms—and this is where the overall picture becomes complicated.

With the assembling of the thirty-sixth Congress in December 1859, everything pointed in the direction of a general revision of the overland service. Partisans of the central route were active, but they met with considerable opposition. California wanted direct mail service over the shorter central route, and the Pony Express was seen as a way to focus national attention on the route, thereby shifting funding away from Butterfield's longer but successful route. So, it was with the idea of demonstrating the practicability of the central route for year-round travel and to secure an enlarged mail contract that the Pony Express scheme was conceived. And William H. Russell decided he was the man for the job.

Prior to 1859, Russell owned Leavenworth and Pike's Peak Express company, a stage line that linked Missouri and Kansas with Denver and an

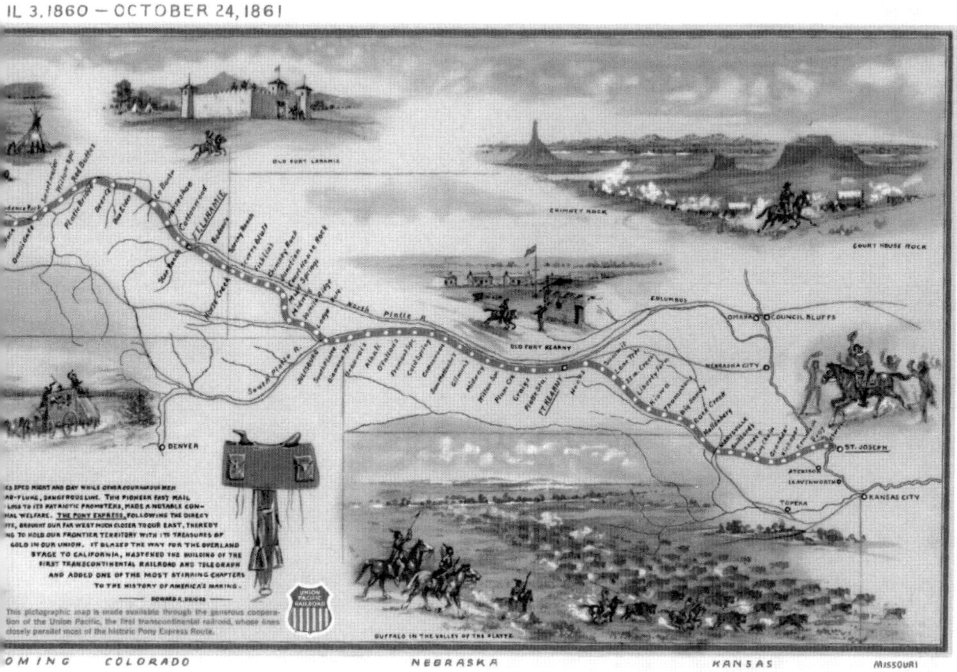

extended mail route to Salt Lake City. When the company went bankrupt in 1859, he turned to his partners in his other still-flourishing freight company for a loan. This company, the Central Overland and Pike's Peak Express Company, was plagued with thievery among its employees. It took a cold-eyed gunman, Jack Slade, to put a stop to it.

Although Russell's stage line continued to lose money, he was determined to win a monopoly on the government mail contracts in the Rocky Mountains. This was bid as a dual stagecoach and pony express service, and the government contracted $1 million for the operations. The agreement specified that the obligation would end when the telegraph line was completed.

Russell did not receive the mail contract. Instead, the law provided that the Butterfield Overland Mail line be moved to the central route. However, a working arrangement was made whereby Russell, with his partners Alexander Majors and William Waddell, would operate the Pony Express and the daily mail coach service from the Missouri River to Salt Lake City, while the Overland Mail Company would operate the portion of the line west of Salt Lake City.

The Pony Express was inaugurated as an advertiser and demonstrator of the central route rather than as an immediate moneymaking scheme. From the standpoint of the nation, the Pony Express was a success by demonstrating the practicability of the central route and marking the path for the first transcontinental railroad. By shortening the distance between the Atlantic and Pacific coasts, it helped unite the Pacific and the Rocky Mountain with the Union—especially during the beginning of the Civil War.

On January 27, 1860, Russell announced the formation of the Pony Express service from St. Joseph, Missouri, to Sacramento, California, commencing on April 3, 1860.

The following wire dispatch is the first definite notice regarding the Pony Express:

> *Washington, January 27, 1860*
> *To John W. Russell: Have determined to establish a Pony Express to Sacramento, California, commencing the 3rd of April. Time ten days.*
> *—Wm. H. Russell*

The *San Francisco Bulletin* of February 29, 1860, reported that a February 6 St. Louis newsletter had announced that the "horse express" would begin operations on April 3 and that "horses and riders are now being placed on the line ready for this new enterprise in this fast age."

Many stations were already established from the eastern terminus to Salt Lake City, but beyond that, more extensive preparations had to be made. Horses were purchased and distributed along the line. The best riders obtainable were stationed along the route, as were the relay and extra animals. Russell, Majors and Waddell organized and put together the Pony Express. The undertaking involved acquiring 120 riders, 184 stations, 400 horses and several hundred personnel during January and February 1860. Within two months, they were ready.

True to plans, on April 3, all was ready for a simultaneous start from the two ends of the line. In San Francisco, a "clean-limbed, hardy little nankeen-colored pony" stood waiting for his gallop into history. This pony had but a short run to the boat, which would carry the express to Sacramento—the real start of the Pony Express.

That same day, at St. Joseph, a large crowd assembled to witness the launching of the history-making Pony Express. Resembling Butterfield's auspicious start, this express had to wait for the Hannibal and St. Joseph train to arrive with the connecting mail. It was hours late. But when it did chug into town, mail was quickly shifted into bags lettered "Overland Pony Express," and they were off in a matter of moments.

Unlike Butterfield's start, receptions of the ponies along the route and at opposite ends of the line were quite enthusiastic. Because of the telegraph, people ahead of the ponies knew to be waiting. Crowds gathered and cheered as the ponies flew past. Ceremonies of reception, music, "jollification" and some speeches followed the Pony's arrival. Bonfires were kindled, and bands played well into the night. Surely, John Butterfield was envious.

The route taken was the existing overland mail route via Salt Lake City and generally followed the trail made by Mormon pioneers in 1847 and the California Argonauts of 1849. It also used an old overland military road in Missouri. Along the route, stations were made of locally available material and established about fifteen miles apart (the distance a horse could manage easily).

The four-hundred-plus horses used were the best in the West, handpicked by co-founder Alexander Majors. Half-breed California mustangs, averaging fourteen and a half hands high and 900 pounds, composed most of the herd. These horses were famous for their speed, endurance, dependability and cost. Majors paid an average of $200 (about $5,400 today) each. The ponies were fed and housed carefully, as they faced the severe test of running ten, fifteen or twenty-five miles with few breaks.

Riders had to be just as tough. The pick of the frontier, these young men were selected for their nerve, light weight (they could not weigh over 120 pounds) and general fitness. They were armed but generally depended on the fleetness of their horse to get them out of trouble. The rider's life was exciting and dangerous. At the beginning, each rider rode between thirty and fifty miles using three horses. Later, the run was extended from seventy-five to one hundred miles. A round trip over this run was made twice a week, for which the rider received a salary between $50 and $150 a month ($1,350 to $4,050 today).

The mail was carried in four small leather bags called cantinas, each about six by twelve inches in size. These were sewed to a mochila, a leather saddlebag suspended in such a way that one letter bag was in front and one behind each leg of the rider. The same mochila was transferred from pony to pony and rider to rider until reaching its destination.

The maximum weight for any mail was twenty pounds. The initial charge was five dollars per each half ounce but was gradually reduced to one dollar per half ounce. In today's currency, that equates to about twenty-seven dollars per half ounce. Additionally, each letter had to pay the regular ten-cent government postage. Needless to say, only correspondence of extreme importance was carried by the Pony Express.

The approximately 1,900-mile route roughly followed the Oregon and California Trails to Fort Bridger in Wyoming and then the Mormon Trail (known as the Hastings Cutoff) to Salt Lake City, Utah. From there, it followed the Central Nevada Route to Carson City, Nevada, before passing over the Sierras into Sacramento, California.

Following is the waybill of the notable first trip of the Pony Express:

Left St. Joseph 6:30 pm, April 3
Arrived Salt Lake City 6:30 pm, April 9
Arrived Carson City, Nevada 2:30 pm, April 12
Arrived Placerville 2 pm, April 13
Arrived Sacramento 5:30 pm, April 13
Arrived San Francisco 1 am, April 14

As the telegraph neared completion, interest waned in the Pony Express, except for the portion between the approaching termini of the telegraph line. As telegraph lines from both coasts advanced toward each other, the Pony Express runs became shorter and shorter. When the line was completed on October 24, 1861, the Pony Express came to a close.

Chapter 30
TELEGRAPH

The telegraph itself did not spell the end of the Overland Mail Company, but it certainly didn't help matters. No longer was it necessary to rush important news twenty-five days and 2,800 miles across a hostile environment. Now a single wire did the job. Just as Butterfield knew the train would someday force him out of business, he knew the telegraph would do the same. The telegraph allowed instant communication over vast distances for the first time in human history.

The idea behind the telegraph—sending electric signals across wires—originated in the early 1700s, and by 1798, a rough system was being used in France. In 1828, the first telegraph in the United States was invented by Harrison Dyar, who sent electrical sparks through chemically treated paper tape to burn dots and dashes. Further developed in the 1830s and '40s by Samuel Morse, the telegraph revolutionized long-distance communication.

In addition to helping invent the telegraph, Samuel Morse developed a code (which bears his name) that assigned a set of dots and dashes to each letter of the English alphabet and allowed for the simple transmission of complex messages across telegraph lines. In 1844, Morse sent his first telegraph message, which was transmitted from Washington, D.C., to Baltimore, Maryland.

In 1844, Congress witnessed the sending and receiving of messages over part of the telegraph line. Before the line had reached Baltimore, the Whig party held its national convention there and on May 1, 1844, nominated Henry Clay. This news was hand-carried to Annapolis Junction (between

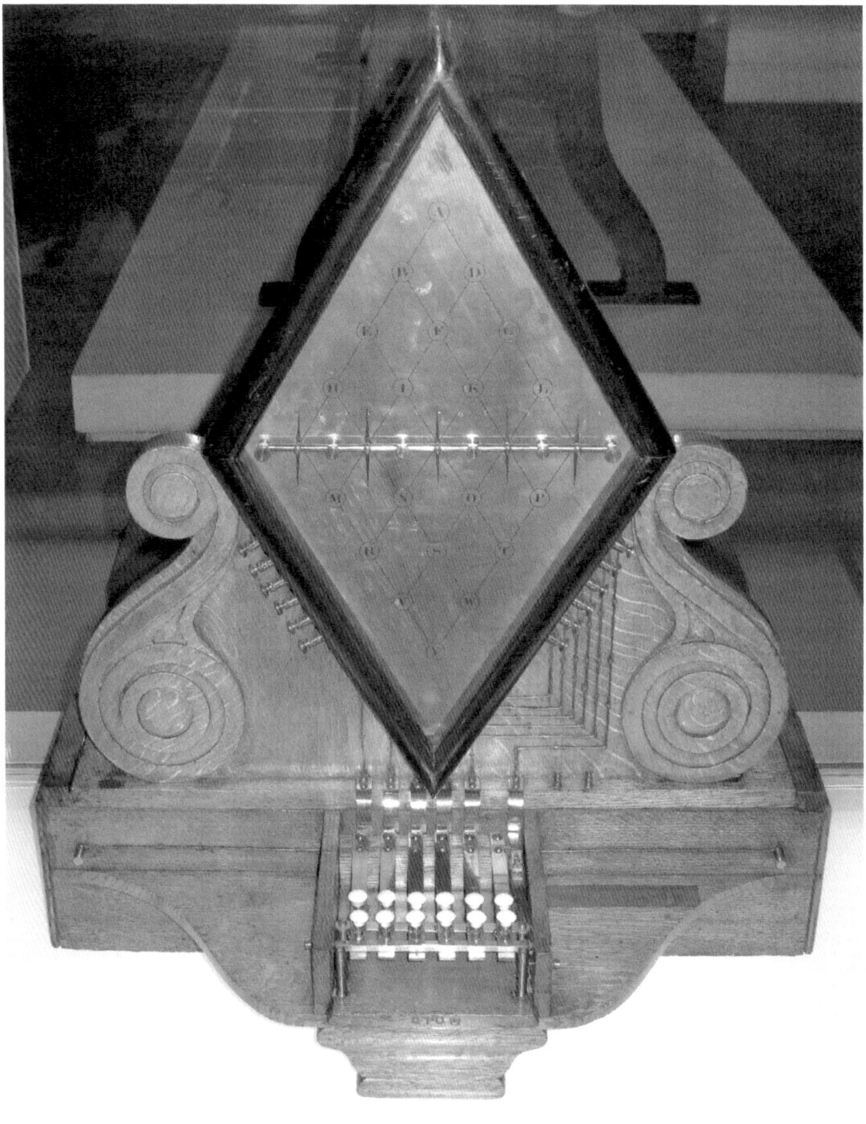

Cooke and Wheatsone's electric telegraph (needle telegraph) dates from 1837 and is now in the London Science Museum. *Photo by Geni William Fothergill.*

Washington and Baltimore), where Morse's partner, Alfred Vail, wired it to the Capitol. This was the first news dispatched by electric telegraph.

Prior to the telegraph, politics and business were constrained by geography. The world was divided into isolated regions. There was limited knowledge of national or international news, and that which was shared was generally quite dated. After the telegraph, the world changed.

The telegraph line strung across the U.S. continent was completed on October 24, 1861. By 1866, a telegraph line had been laid across the Atlantic Ocean from the United States to Europe.

In the newly formed state of California, the telegraph was a source of great curiosity to almost every person along the route, particularly the natives, who looked upon the construction of the line with the greatest wonder. In Arizona, they cut the lines as quickly as they went up. It is reported that many of them, in ignorance of its real purpose and not understanding the use of the poles erected along the road at regular intervals, strung with wire with a cross arm on each pole, conceived the idea and expressed it as their belief that the Yankees were fencing in the country with crosses to keep the devil out.

Sending messages was expensive. In California, the California State Telegraph Company's tariff between San Francisco and San Jose (a distance of fifty miles) was seventy-five cents for ten words and twenty-five cents for every additional five words or fraction thereof. From San Francisco to Stockton, Sacramento and Marysville, the rate was two dollars per ten words. When compared to the Pony Express prices, the telegraph probably seemed a bargain.

Whatever the cost, the telegraph spun the mid-nineteenth-century world on its ear.

Chapter 31
INDIANS

The Butterfield Overland Route passed directly through Indian country. As discussed earlier, in general, the Native Americans left the stagecoaches alone, allowing them to pass east to west as long as they continued moving.

By 1857, Indians in the eastern part of the United States were peaceful. The Choctaws in Indian Territory were not much interested in raising war on the Overland Mail or any other stagecoach company that passed through. The Comanches in Texas, however, were another story.

Expert horsemen, the Comanches became something of the Cossacks of the Plains, making mounted attacks against the American military until the Texas Rangers came along. The Rangers, carrying the .44-caliber Walker-Colt six-gun (Samuel Walker was a Ranger himself), were so unrelenting and fierce that the Mexicans called them "Los Tejanos Diablos." In the 1850s, the Texas Rangers were assigned seek-and-destroy missions. In fact, Texas Ranger Rip Ford tracked Comanches across the Canadian River into their own territory and wiped out two or three Comanche settlements. The reputation of the Rangers had an extremely effective result in changing the Indians' attitudes.

On the other hand, the Apache in Arizona and New Mexico were not skilled horsemen. They much preferred eating horses and mules instead of cultivating them. Since they did not use horses much, their line of defense was entrapment—blocking narrow gaps where stages could not turn around. They would shoot the lead mule and then attack the coach. They seldom

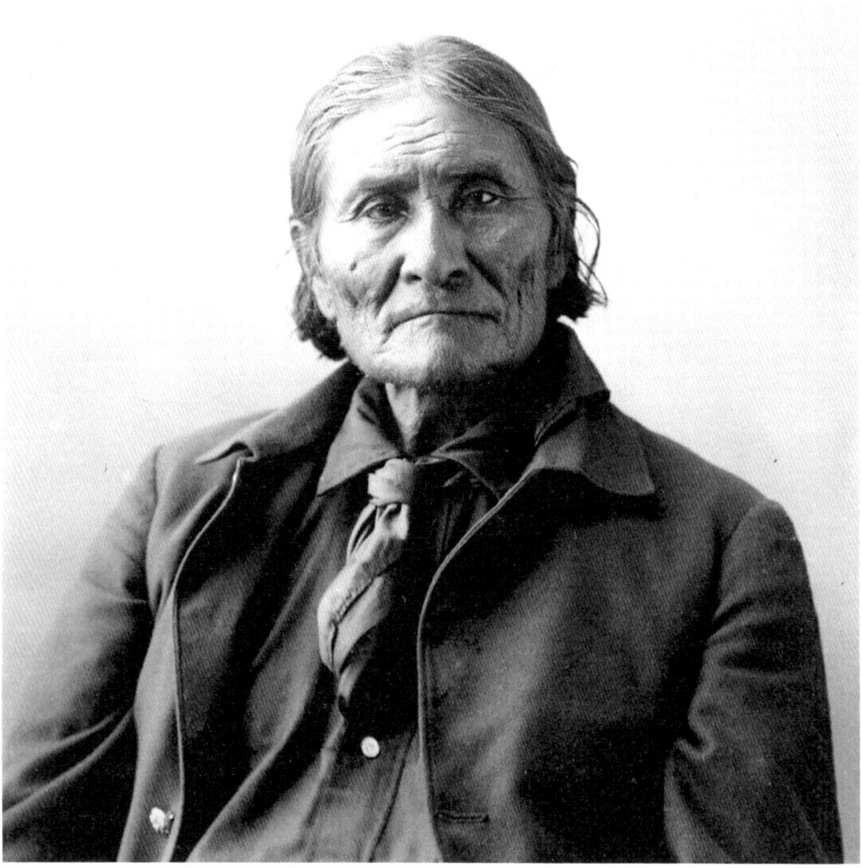

Geronimo (1829–1909), whose name means "one who yawns," was a Bedonkohe Apache leader who fought against Mexico and the United States. *Photo by Frank A. Rinehart, 1898.*

attacked in the open or at night. Apaches dressed only in a clout and high moccasin-type boots that reached well up over their thighs. Traveling light, they carried bows and quivers (later guns), knives and little sacks for food.

As stated previously, John Butterfield firmly believed in peace. He issued about 150 Colt Dragoon six-guns (1849 model) to his employees, but they were instructed to ask questions before shooting. This worked well—until the Bascom Affair.

Cochise, the undisputed chief of the Chiricahuas, was Apache from end to end. He would allow no pictures to be taken of him and believed in honesty above all else. He kept his people on good terms with the Americans and especially the Overland Mail Company. It wasn't until the Bascom Affair that Cochise caused any trouble for that particular stage line. Other stage lines, such as the San Antonio–San Diego, weren't as fortunate or smart enough to work with the Indians.

As a result of the Bascom Affair, however, the Indians decided to take it out on the stage station at Apache Pass and the next through stage. The Indians planned this attack for February 16, 1861. Usually on schedule, the westbound that day was two hours early. It cleared the station and reached the pass while it was still daylight. The stage ran into a heap of dry grass and tumbleweed that had been placed across the trail forming a fire ambush.

Not sensing any danger, the stage crew cleared the brush and went on. In another mile or so, they came upon the remains of burned wagons and the residual of the military and Indian activities of the past two weeks. They were aware of the Bascom Affair but decided to continue on.

However, the eastbound stage had no information as to what had occurred. Occupants of that stage contained A.B. Culver, brother of the killed station keeper at Apache Pass, and eight other passengers. As they entered Apache Pass, shots rang out and two horses went down. The passengers kept up a covering fire while Culver and another man cut the dead animals out of the rig. They fought their way to the safety of the station and holed up. The Indians raided the horse corral without attacking that particular station.

General Thomas C. Devin, in an annual report to Fort Lowell dated January 25, 1869, wrote, "Cochise had earlier guarded the road for the Overland before the attempt to take him prisoner caused him to make war."

And make war they did. They allowed travel through their country only at their pleasure. And most of the time, it didn't please them. In July 1861, as the Confederates marched into La Mesilla, Butterfield employee Emmett Mills and six other Union loyalists headed to California to avoid being imprisoned by the Rebels. On July 21, they rode into an Indian ambush about a mile beyond Cooke's Spring. The combined forces of Cochise and Mangas Coloradas pursued the coach to a nearby hilltop, where a three-day siege took place. All seven white men were killed, but not before several Apaches were killed or wounded.

Henry Skillman, acting as superintendent of the San Diego mail line, intended to "withdraw the stock from the line for the present, and until Indian affairs become quiet in the Territory." The *Mesilla Times* quoted him as saying that "the service on this route will be early resumed." He promised arrivals and departures for Santa Fe, El Paso, San Antonio and California on both a weekly and biweekly basis.

He started off well but was back on August 17. The mail, he reported, "had been overhauled by the U.S. authorities, and the letters opened and read...The Indians captured the mail again on the Jornada and scattered it over the ground. It was collected again, and finally reached here, much of it without envelopes."

Once the Overland pulled out of the southern route, the Indians moved right back in, taking the opportunity to push the white man out. Kiowas and Comanches, as well as Apaches, continued to attack stagecoaches and mail service. It would be twenty-five years before peace would settle over the Southwest.

Chapter 32
THE CIVIL WAR

Nothing good comes out of war, and the Butterfield Overland Mail Company was no exception. Between Butterfield's ill health and departure as president, the Bascom Affair, Indians, the telegraph, the Pony Express and now the Civil War, the notion of mail arriving on both coasts in a timely fashion turned into a memory.

Friction between the North and South had reached its boiling point in late 1860. Congress knew war was imminent. The only reason the mail kept moving was that both sides needed to keep communication open with the Southwest.

By the spring of 1861, Texas had blocked nine hundred miles of the Butterfield trail when it seceded from the Union. Since it was a large part of the road, that move essentially shut down the southern route. Mail stacked up in Tucson as coaches turned around to return to California. On March 6, the last eastbound coach was loaded and sent on its way.

On March 2, 1861, Congress passed a bill that would move the Butterfield Company up to the central route. The Pony Express would service St. Joseph to Salt Lake City, while Butterfield would service Salt Lake City to Sacramento. This move also allowed the OMC's property to stay out of Rebel hands.

Butterfield's Overland Mail Company dream was ended officially.

Logistically, this move meant reorganizing the entire 2,992 miles of line, not just the New Mexico–Texas–Arizona area. The Overland Mail Company had to abandon all the stations, wagon roads, wells and tanks, as well as other improvements made on the southern route.

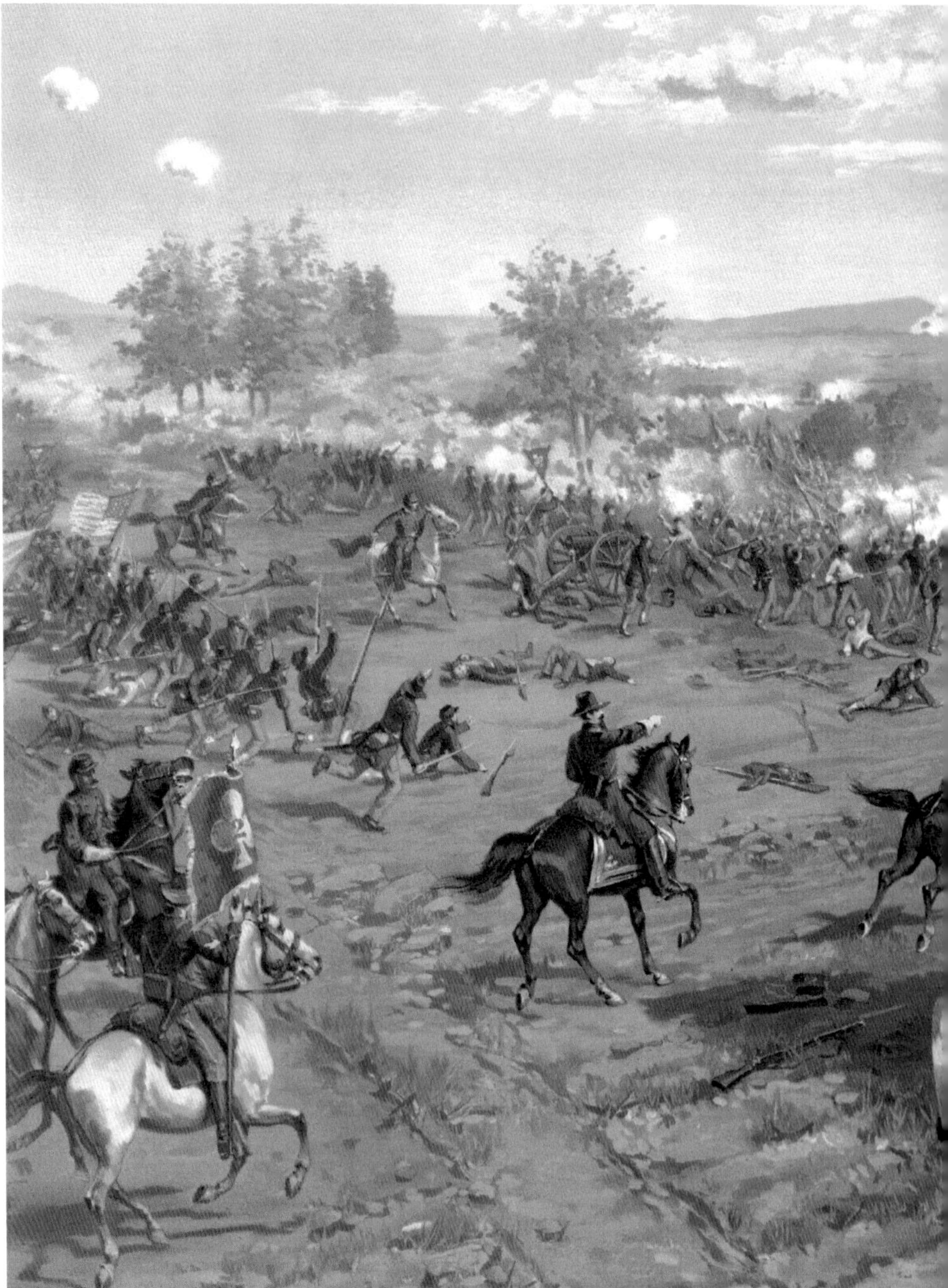

The Civil War rang the death knell for Butterfield Overland Mail. *Public domain.*

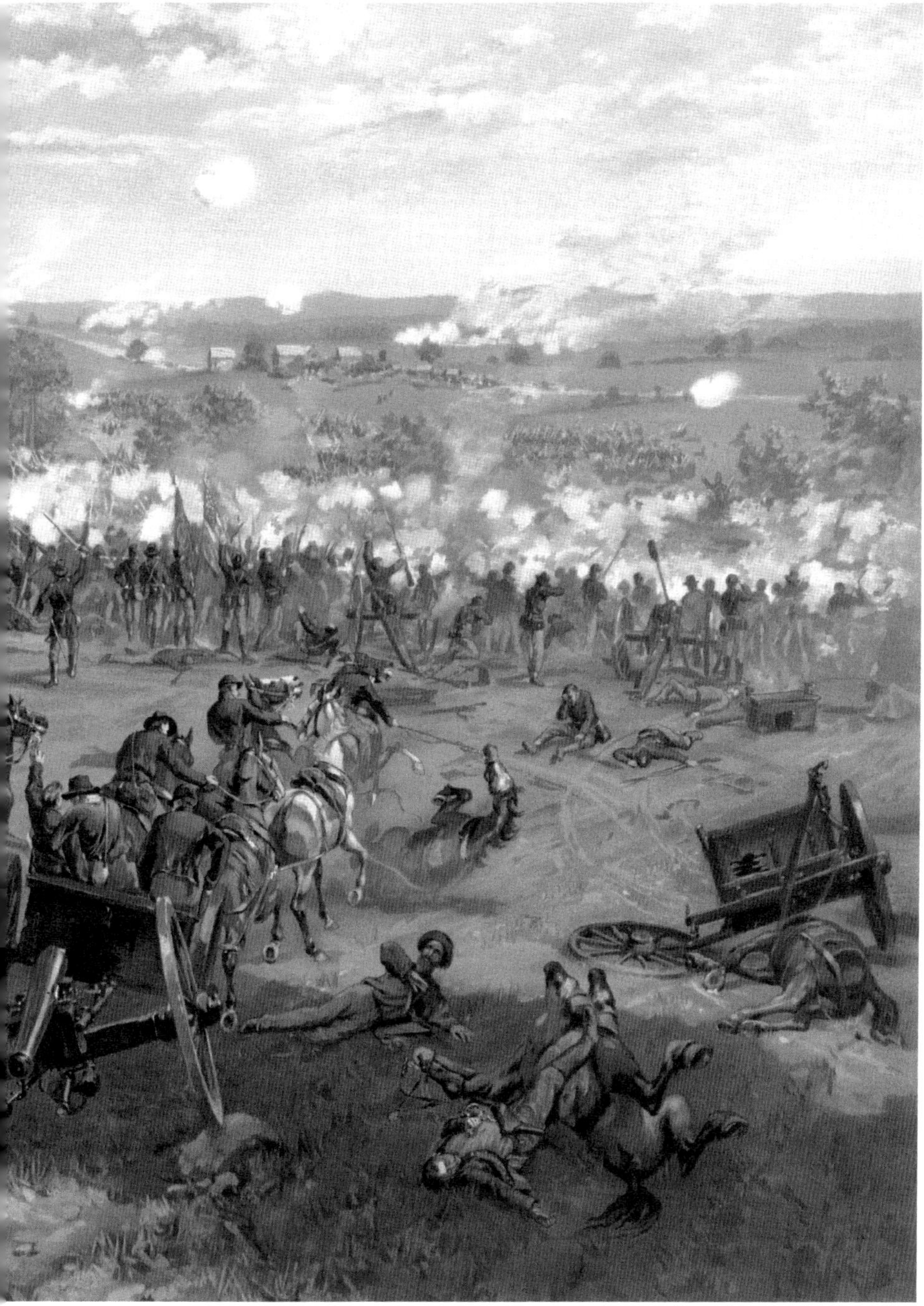

Not willing to abandon a southern route, George Giddings, through political connections with the Confederates, obtained the new government's blessing to keep the mails moving over the Butterfield trail. Giddings, along with Robert E. Doyle and John T. Doyle, organized a new company—the Overland Mail Line. In addition to Confederate approval, they also received a Union government contract to move the mail from San Antonio, Texas, to all points in California. This worked well until July of that year, when Texas invaded New Mexico. The central route would have to do.

Not everyone was happy with the change. People in California still wanted their mail. For a while, they got it. Newspaper mail was usually carried through in twenty-eight days and the letter mail in twenty. Then winter came, and the Colorado Rockies and Sierra Nevadas were slammed with record-breaking snowfall. Proponents of the central route blamed the poor service on bad weather, not the route choice, refusing to believe it was not a year-round route.

The champions of the southern route quickly declared the service a failure. A *Los Angeles Star* reporter writing in the December 11, 1861 issue said, "The Daily Overland Mail is a failure—a gigantic humbug and swindle—and yet the papers up here [in Sacramento] will not say anything upon the subject. While the Southern route was a complete success, that same miserable sectional jealousy that has brought ruin upon our country, would not let it live, because it happened to pass south of Mason and Dixon's line…If the Mail Company cannot perform the service let the steamers bring us our mail."

It didn't get better. Between the snow and the mud, the stagecoaches had a difficult time. In the mountains, the roads drifted full of snow, and on occasion, the teams got off track and were lost. Deep, stiff mud made the coaches too heavy for the team in some places, causing the passengers to get out and walk. Mailbags full of government documents were frequently employed to fill in chuckholes in the worst patches of the road.

When the Civil War broke out, New Mexico Territory succumbed to the wiles of Confederate lieutenant John R. Baylor and the urgings of Arizona citizens to merge with the Confederate State of Arizona. The line was drawn east–west, with New Mexico in the northern half. Since most of Butterfield's personnel were from the northeast, it seemed logical they would side with the North. Besides, the OMC's east and west termini were both northern. In Confederate Arizona, most, if not all, of the Rebel soldiers were actually from Texas and took orders from there. When they were defeated thirteen months later, they retreated to Texas. New Mexico became New Mexico again, and the boundary changed to the present-day north–south line.

With the secession of Texas, the great bend of the Oxbow Route was wiped out. In fact, the last stage east ran through some Confederate soldiers. After the South had closed down the original Butterfield route, there was no regular scheduled service across America's midsection. The southern route had proved to be, and still is, the best year-round cross-country route.

The Butterfield Overland Mail Company came to a screeching end. The last mail west left Fort Smith in April 1861 but did not make it through—it was held up at Fort Chadbourne.

The mail at Tucson was the largest accumulation, and the last mail was loaded and left town on March 6, 1861. This stage cleared Franklin/El Paso on March 9, 1861, and headed out. About thirty-five miles southwest of Fort Chadbourne, the stage overtook a party of Texas volunteers on the trail. The troops parted to let the stage through. All was going well until the conductor, out of ignorance or habit, blew his bugle and loudly called out, "Make way for the United States mail!"

It was not a good idea, considering they were in Confederate territory. The Texans laughed and cursed while ground-looping their horses and rapidly dismounting. They promptly seized the wheels of the stage and shouted to the driver, "Go ahead—let's see you make way."

The driver let out the horses and popped his whip while the passengers, fearing the worst, remained absolutely quiet and, as one stated later, "sat on nervous seats." Try as he might, the driver could only get the team to drag the stage a few feet, and the Tejanos held back the wheels while they cajoled the driver with curses and more raucous laughter. After fifteen or twenty minutes of this horseplay, the Tejanos finally released the wheels, and amid catcalls, jeers and more profanity, the coach shot ahead into history.

Thus ended the Overland Mail.

EPILOGUE

Butterfield's company was officially named the "Overland Mail Company," but in Missouri, it was known as the "Great Southern Overland," and in California, it was the "California Overland Express." In Arizona, New Mexico and Texas, it was simply "The Butterfield." There were also other "overland" mail lines, adding to the confusion. Whatever you call it, there was only one Overland Mail Company and only one John W. Butterfield. I still visit La Mesilla every chance I get. I walk the streets and hear the rumble of stagecoach wheels and the jehu hollering a greeting as he brakes at the Corn Exchange. More than once, I stroll past and eventually enter the headquarters of the Butterfield Overland Mail Company. The buildings are now home to tourist stores and wine shops, but I can still see and hear the men who stitched our country together.

RULES OF TRAVEL: STAGECOACH DECORUM

- When a driver asked a passenger to get out and walk, one was advised to do so and not grumble about it.
- If the team of horses ran away, it was better to sit in the coach because most passengers who jumped out were seriously injured.
- Smoking and spitting on the leeward side of the coach was discouraged.
- Drinking spirits was allowed, but passengers were expected to share.
- Swearing was not allowed, nor was sleeping on your neighbor's shoulder.
- Travelers were discouraged from pointing out spots where murders had occurred, especially when "delicate" passengers were aboard.
- Greasing one's hair was discouraged because dust would stick to it.

TRAVEL COMPARISON: 150 YEARS AGO VERSUS TODAY

THEN: Passengers were limited to forty pounds of luggage.
NOW: Airline passengers are limited to forty pounds of luggage.

THEN: Breakfast cost one dollar.
NOW: A breakfast burrito at McDonald's runs about one dollar.

THEN: One-way fare was $150—in gold.
NOW: Bargain round-trip airfare is $150—in plastic.

THEN: Cramped, crowded seats, knees up around your chin, packages beneath your feet.
NOW: Fly anywhere in coach.

LIST OF FIRSTS

1. Butterfield created stage stations, much like today's "rest areas" and "truck stops."
2. As a result of a surveying party, the town of Yuma, Arizona, got its auspicious start.
3. El Paso's American growth originated with the Overland Mail.
4. Los Angeles grew after connecting with the world via the Overland Mail.
5. The Overland became the longest stagecoach line in the world under one management.
6. Butterfield organized and ran the first truly transcontinental traffic line on the continent.
7. A new concept, strict adherence to the schedule for the entire period, proved to be noteworthy.
8. For the first time, drivers were assigned to small sections of the route while mail and passengers went on through.
9. Butterfield designed wide tires to maintain stability under rough conditions.
10. Even compared to smaller companies, Butterfield's safety record stood unequaled.

LIST OF WAY STATIONS

First Division	Miles
San Francisco to Clark's	12
San Mateo	9
Redwood City	9
Mountain View	12
San Jose	11
Seventeen Mile House	17
Gilroy	13
Pacheco Pass	18
St. Louis Ranch	17
Lone Willow	18
Temple's Ranch	13
Firebaugh's Ferry	12
Fresno City	19
Elk Horn Spring	22
Whitmore's Ferry	17
Cross Creek	12

First Division	Miles
Visalia	12
Packwood	12
Pule River	14
Fountain Spring	14
Mountain Home	12
Posey Creek	15
Gordon's Ferry	10
Kern River Slough	12
Sink of Tejon	14
Fort Tejun	16
Reed's	8
French John's	14
Widow Smith's	24
King's	10
Hart's	12
San Fernando Mission	8
Cahuengo	12
Los Angeles	12

Total: 80 hours, 462 miles

Second Division	Miles
Los Angeles to Monte	13
San Jose	12
Chino Ranch	12
Temascal	20
Laguna Grande	10
Temecula	21
Tejungo	14

Second Division	Miles
Oak Grove	12
Warner's Ranch	10
San Phelipe	16
Vallecito	18
Palm Springs	9
Corrizo Creek	9
Indian Wells	32
Alamo Mocho	24
Cook's Well	22
Pilot Knob	18
Fort Yuma	10

Total: 72 hours and 20 minutes, 282 miles

Third Division	Miles
Fort Yuma to Swiveller's Ranch	20
Filibuster Camp	18
Peterman's	19
Griswell's	12
Flap Jack Ranch	15
Pointer Mountain	22
Oatman Flats	20
Tucson	28
Murderer's Grave	20
Gila Ranch	17
Maricopa Wells	40
Sacaton	22
Picacho del Tucson	37

Total: 71 hours and 45 minutes, 280 miles

FOURTH DIVISION	MILES
Tucson to Seneca Springs	35
San Pedro River	24
Dragoon Springs	23
Apache Pass	40
Stein's Peak	35
Soldier's Farewell	42
Ojo de la Waca	14
Mimbres River	16
Cook's Springs	18
Picacho	52
Fort Fillmore	14
Mesilla	25
Franklin/El Paso	22

Total: 82 hours, 360 miles

FIFTH DIVISION	MILES
Franklin to Hueco Tanks	30
Cornudod de los Alamosos	36
Pinery	56
Delaware Springs	24
Pope's Camp	40
Emigrant Crossing	65
Horse Head Crossing (no station)	55
Head of Concho	70
Concho Creek	30
Grape Creek	23
Fort Chadbourne	30

Total: 126 hours and 30 minutes, 458 miles

Sixth Division	Miles
Fort Chadbourne to Station No. 1	12
Mountain Pass	16
Smith's	12
Phanton Hill	30
Clear Fork of the Brazos	26
Francis's	13
Ft. Belknap	22
Murphy's	16
Jacksboro	19
Earhart's	16
Conolly's	16
Davidson's	24
Gainsville	17
Sherman	15
Colbert's Ferry	13.5

Total: 65 hours and 26 minutes, 282.5 miles

Seventh Division	Miles
Colbert's Ferry to Fisher's	13
Nale's	14
Boggy Depot	17
Gary's	16
Waddell's	15
Blackburn's	16
Pusley's	17
Riddell's	16
Rolloway's	18
Trayon's	19

Seventh Division	Miles
Walker's	16
Fort Smith	15

Total: 38 hours, 192 miles

Eighth Division	Miles
Ft. Smith to Woosley's	16
Brody's	12
Park's	20
Fayetteville	14
Fitzgerald's (station #2)	12
Callahan's	22
Harvin's	10
Crouch's	16
Smith's	15
Ashmore's	20
Springfield	13
Evan's	9
Smith's	11
Bolivar	11.5
Yost's	16
Quincy	16
Bailey's	10
Warsaw	11
Burn's	15
Mullhallen's	20
Shakelford's	13
Tipton	7

Total: 48 hours and 55 minutes, 318.5 miles

Ninth Division	Miles
Tipton to St. Louis (by railroad)	160

Total: 11 hours and 45 minutes, 160 miles

Overall Total: 24 days, 18 hours and 26 minutes; 2,795 miles

TIMELINE

1846—President Polk starts the War with Mexico

1846—Oregon country obtained via treaty with Great Britain, starting the great northwest migration

1846—Stephen Kearney and the Army of the West occupy New Mexico; Cook begins building southern wagon road to California

1847—Treaty of Guadalupe Hidalgo ends the war with Mexico with the cession by Mexico of most of California, Arizona, New Mexico, Utah, Nevada and Colorado

1848—Gold is discovered in California

1849—California migration explodes—the gold rush is on!

1849—Oregon Territory organized

1850—California gains statehood

1854—Gadsden Purchase restores part of southern New Mexico and southern Arizona left out by error in the Treaty of Guadalupe Hidalgo

1856—San Antonio–San Diego Mail Company given contract to carry U.S. mail from San Antonio to San Diego

1857—John Butterfield given government contract to carry mail from St. Louis to San Francisco; Butterfield builds and stocks stations, grades roads, hires men and develops watering stations

1858—Butterfield's Overland Mail Company inaugurates first runs with coaches leaving from San Francisco and Tipton, Missouri, in September

1859—Oregon gains statehood

1860—Congress establishes the Territory of Colorado and annexes all of New Mexico above the thirty-seventh parallel to Colorado

1861—The Civil War blocks the Butterfield southern route to California; all Union mail contracts move through Northern states to California

TERMS AND SLANG

boot: A deep luggage carrier at the rear of a stagecoach and also under the driver's seat.

box: The stage driver's seat.

brother whip: The stagecoach driver, also called simply "whip."

bull-whacker: A driver of a freight wagon, usually pulled by oxen.

Charlie: A stagecoach driver.

conductor: The person who rode with the driver of a stagecoach and collected fares, took care of passengers and was responsible for the mail.

corduroy road: A road created by logs laid across a swampy, low-lying area and placed together or "ribbed" like corduroy cloth—not to be confused with a plank road.

division: The 250 miles of trail on the Overland Route belonging to a division agent.

division agent: The person in charge of 250 miles of trail on the Overland Route, also called a superintendent. The division agent was in charge of purchasing animals and equipment, maintaining the stations and hiring station keepers, conductors, hostlers, drivers and blacksmiths.

drag: An object attached to the back of a stagecoach to slow it when going downhill.

express: Valuables other than mail and newspapers such as small packages, payroll, and gold and silver bars.

expressman: A messenger carrying express items.

groom: A stableman; one who takes care of the horses.

hame: One of two curved bars fitted to a horse collar that hold the traces of a harness.

hangers-on: A phrase used by the more well-off riders inside the stagecoach to refer to those who were riding on top.

home station: A larger station where meals and lodging were provided to passengers.

hostler: Someone who took care of the horses and mules at the stage stops and also changed the team.

jehu: A stagecoach driver, possibly taken from the name of a Biblical character who drove fast and furiously.

johnnycake: A type of cornbread often provided to travelers at stage stops.

leaders: The two front animals of the team. These would be the smallest, smartest and most alert of the team.

lines: Another term for reins.

near: Term used to refer to the horses or mules on the left side of the team, farthest from the driver.

off: Term used to refer to the horses or mules on the right side of the team, nearest the driver.

overland: Across land rather than by water.

postrider: A person who carried the post, or mail, on horseback.

reaches: Bars that connected the rear axles with the forward part of the coach.

reinsman: A stagecoach driver.

relay: A team of horses or mules kept in readiness at a way station to relieve the team of an approaching stagecoach.

ribbons: Another term for reins.

rig: A harness.

road agent: A stagecoach robber.

run: The distance between stage stations.

shotgun: A stagecoach guard; also referred to as a "shotgun messenger."

singletree: A horizontal crossbar, to the ends of which the traces of a harness are attached.

stage horn: A horn or bugle blown by the conductor as the stage neared the station.

stage line: A stagecoach company.

stagers: The men who ran or owned a stage business.

staging: The business of carrying people and mail by stagecoach.

station: The place at which a stagecoach stopped.

station keeper: The person in charge of the station stop.

superintendent: The person in charge of 250 miles of trail on the Overland Route, also called a division agent.

swingers: If a team had six animals, the two in the middle were called "swingers" and were easily controlled by the other two pairs as they did their work leading or stabilizing.

swing station: A small stage station where the team was changed. Usually consisting of little more than a small cabin and barn or corral, the swing station was typically manned by just a few stock tenders. Stages stopped at swing stations for about ten minutes before moving on.

team: The horses or mules pulling the stagecoach, usually consisting of four to six animals.

thorough brace: A multilayered leather strap that supported the body of the stagecoach.

traces: The side straps by which a horse pulled the stagecoach.

transcontinental: Across the United States.

turnpike: A main road on which travelers paid a toll or fee.

waybills: Stage line advertisements posted in towns and villages.

way station: see station.

wheelers: The two animals at the rear of the team. These would be the largest and strongest animals to provide stability nearest the coach.

wheelwright: A person who makes or fixes wheels.

whip: The stagecoach driver, also called "brother whip."

whippletree: The horizontal bar at the front of the stagecoach, to which singletrees are attached.

BIBLIOGRAPHY

Austerman, Wayne R. *Sharps Rifles and Spanish Mules: The San Antonio–El Paso Mail, 1851–1881.* College Station: Texas A&M University Press, 2000.

Fradken, Philip L. *Stagecoach: Wells Fargo and the American West.* New York: Free Press, 2003.

Griggs, George W. *History of Mesilla Valley; or, The Gadsden Purchase.* Las Cruces: Bronson Printing Company, New Mexico, 1930.

Hackler, George. *The Butterfield Trail in New Mexico.* Las Cruces, NM: Yucca Enterprises, 2005.

Hafen, LeRoy R. *The Overland Mail: 1849–1869.* Norman: University of Oklahoma Press, 2004.

Kiser, William S. *Dragoons in Apacheland: Conquest and Resistance in Southern New Mexico, 1846–1861.* Norman: University of Oklahoma Press, 2013.

McFie, Maude Elizabeth. *A History of the Mesilla Valley to 1903.* Las Cruces, NM: Yucca Tree Press, 1999.

Ormsby, Waterman L. *The Butterfield Overland Mail: Only Through Passenger on the First Westbound Stage.* Berkeley: University of California Press, 2007.

Thomas, David G. *La Posta*. N.p.: Doc45 Publishing, 2013.

Tompkins, G.C. *A Compendium of the Overland Mail Company on the Southern Route, 1858–1861*. El Paso, TX: Talina Corporation, 1985.

INDEX

A

American Express 42
anvil salute 91, 119
Apache Pass 115, 133
Arizona 17, 20, 39, 51, 56, 59, 62, 63, 65, 67, 68, 80, 84, 112, 114, 115, 132, 133, 140, 149, 151, 155, 158, 161, 167
Austin, Stephen F. 21

B

Baltimore and Ohio Railroad 35
Bascom Affair 152
Bascom, George 134, 139
Bean, Roy 63, 114
Bean, Sam 63, 114
Birch, James 38, 51
Boston Mountain Drive 63
Brown, Aaron V. 39, 47, 51
Bugscuffle Road 98
Butterfield, Daniel 90
Butterfield, John, Jr. 95
Butterfield, John W. 41, 56, 62, 68, 80, 89, 129

C

California 10, 16, 17, 20, 21, 23, 25, 26, 27, 29, 30, 33, 34, 38, 43, 46, 49, 50, 51, 55, 57, 62, 63, 65, 67, 71, 72, 89, 90, 91, 117, 118, 132, 141, 142, 144, 145, 146, 149, 153, 155, 158, 161
Celerity stagecoaches 73, 98, 120
Chiricahuas 68, 133
Choctaws 45, 99
Cochise 135, 140, 152
Colbert's Ferry 45, 103
Comanches 151
Concord stagecoaches 71, 98, 120
Corn Exchange Hotel 114
Culver, Charles 135

D

Davis, Jefferson 17
Dinsmore, William 43, 131
Dragoon Springs 62

E

El Paso 10, 18, 20, 23, 38, 46, 51, 55, 83, 109, 110, 111, 125, 153, 159

F

fare 93
Fayetteville, Arkansas 62, 98
Fort Chadbourne 45
Fort Fillmore 111
Fort Phantom Hill 106
Fort Smith, Arkansas 20, 38, 45, 47, 99
Fort Yuma 117
Franklin, Texas 109
Free, Mickey 139
Frémont, John C. 26

G

Gadsden, James 17
Gadsden Purchase 18, 34, 36, 56, 112
Gilroy, California 120

H

horses 85
Houston, Sam 21

J

jehu 79, 80

L

La Mesilla 20, 46, 63, 84, 112
La Posta Restaurant 63, 114
López de Santa Anna, Antonio 18, 21
Los Angeles 29, 118

M

Manifest Destiny 21
meals 61
Memphis 47, 48
Mexican-American War 17
mules 85

N

New Mexico Territory 9, 10, 17, 20, 30, 36, 37, 49, 51, 56, 59, 62, 67, 83, 84, 107, 111, 112, 114, 115, 139, 140, 151, 155, 158, 161

O

Oklahoma 23, 99
Old Wire Road 98
Ormsby, Waterman L. 73, 89
Overland Mail Company 43, 45
Oxbow Route 159

P

Panama 27
Picacho Peak 46, 114, 116
Pony Express 141, 155

R

Rio Grande 29
Rio Grande Valley 109

S

San Antonio–San Diego Line 49
San Francisco 20, 47, 120
Santa Fe Trail 34
Sherman 23, 56, 103
Skullyville 45
Springfield, Missouri 37, 45, 98
stations, home 57
stations, stage 59
stations, swing 57, 61
stations, way 57
Stein's Peak 115
Sutter's Mill 27

T

telegraph 147
Temecula, California 118
Texas 10, 16, 17, 20, 21, 23, 30, 33, 38, 39, 45, 46, 49, 51, 52, 55, 56, 57, 101, 103, 110, 114, 117, 151, 155, 158, 159, 161
Tipton, Missouri 45, 90
Treaty of Guadalupe Hidalgo 17, 26, 37
Tucson 46, 116, 159

V

Van Buren, Arkansas 99
Visalia 65

W

Wells Fargo 42, 131

Y

Yuma 63, 65

ABOUT THE AUTHOR

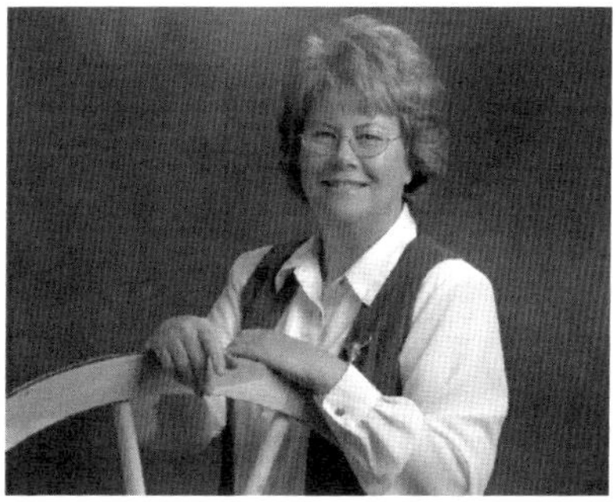

New Mexico native Melody Groves has a deep love for anything cowboy and Old West. Winner of six first-place writing awards, Melody is a member of Western Writers of America, SouthWest Writers and New Mexico Press Women. She writes for *Wild West*, *American Cowboy*, *True West*, *New Mexico Magazine* and other regional publications.

Melody is author of the *Colton Family Saga* series, *Border Ambush* (finalist, 2009 New Mexico Book Awards), *Sonoran Rage*, *Arizona War* (winner, 2008 New Mexico Book Awards) and *Kansas Bleeds*. Also, her essay about living in New Mexico was published in *Voices of New Mexico*. She is a contributing editor for *Round Up* magazine for Western Writers of America, as well as a contributing writer for *American Westward Expansion*, a collegiate history encyclopedia. Her nonfiction book, *Hoist a Cold One! Historic Bars of the Southwest*, looks at the history and beauty of twenty-six front and back bars and was a winner at the 2011 New Mexico Book Awards. *Ropes, Reins, and Rawhide: All About Rodeo* is a guide to understanding the sport.

When not writing, Melody is busy playing rhythm guitar with the Jammy Time Band. She and her husband, Myke, live in Albuquerque, New Mexico.